BENEATH COLD SEAS

BENEATH COLD SEAS

Exploring Cold-Temperate Waters of North America

Photographs by Jeffrey L. Rotman · Text by Barry W. Allen

VNR VAN NOSTRAND REINHOLD COMPANY

NEW YORK   CINCINNATI   TORONTO   LONDON   MELBOURNE

"Do not go gentle into that good night" by Dylan Thomas, copyright ©
1952 by Dylan Thomas, reprinted by permission of New Directions
Publishing Corporation (New York) and David Higham Associates,
Limited (London).

Printed in Hong Kong
Designed by Paul Chevannes

Published by Van Nostrand Reinhold Company Inc.
135 West 50th Street
New York, NY 10020

Van Nostrand Reinhold Publishers
1410 Birchmount Road
Scarborough, Ontario M1P 2E7, Canada

Van Nostrand Reinhold Australia Pty. Ltd.
480 Latrobe Street
Melbourne, Victoria 3000, Australia

Van Nostrand Reinhold Company Limited
Molly Millars Lane
Wokingham, Berkshire, England RG11 2PY

16 15 14 13 12 11 10 9 8 7 6 5 4 3 2 1

Library of Congress Cataloging in Publication Data

Rotman, Jeffrey L.
   Beneath cold seas.

   Includes index.
   1. Marine biology—Atlantic coast (United States)
2. Marine biology—Pacific coast (United States and
Canada) I. Allen, Barry. II. Title.
QH92.2.R67   574.97   81-19850
ISBN 0-442-27058-5   AACR2

For Nancy and Galit

Woe to him that is alone when he falls
and has not another to lift him up.

Ecclesiastes 4:10

CONTENTS

PREFACE

IT WAS a chance meeting on a rainy Sunday evening, in the parking lot of a large New England diveshop. Each of us had heard of the other, but only in the most general terms. One of us (J.L.R.) had been working steadily to capture cold-water marine life on film. The other (B.W.A.) was just beginning to put two decades of cold-water diving experience into words. It seemed a natural collaboration—pictures and words. Perhaps a book on cold-water marine life would be welcomed by all the divers who shared this special experience. Perhaps others who did not know what rich surprises lay beneath cold waters would also take pleasure in it. It was not our intention to create a textbook of marine life, nor did we envision a comprehensive travel guide to the vast cold-water coastlines of North America. We were after a happy marriage of words and photographs that would convey some of those qualities that make the world beneath cold waters so special to both of us and to many, many others.

In fact, however, this work had begun long before our first, chance meeting. J.L.R. had a lifelong interest in natural history, and was particularly fascinated with animal behavior. This led him to marine biology and then to diving and underwater photography. B.W.A.'s passion was diving—penetrating and exploring beneath the water. It had been virtually a compulsion for as long as he could remember, and had led him to the full-time teaching of diving and later to share his experiences more broadly, through writing.

This was how we came to work together on this book. What followed was truly a collective effort, shaped and reshaped by the give-and-take that characterizes all good partnerships. Each felt strongly that the other was essential to the final synthesis and left the order of authorship to a toss of the coin.

We have never forgotten that others have helped us, directly or indirectly, with this project. We are particularly fortunate to have an understanding and creative publisher in Van Nostrand Reinhold. We also owe special thanks to Galit and Nancy (wives of J.L.R. and B.W.A. respectively), who at various times have swum beside us and who have always stood by us; to Ken Beck, the friend and diver who got J.L.R. to take his first plunge; to my coach, Mel Scott, who always has time; to Roy Hauser, skipper of *Truth*, without whose help neither the photography nor the writing on California's underwater environment would be complete; and to Walt Amidon, John Hardy, Dave McLaren, Ned Van Valkenburgh, John Williams, and Jim Willoughby, each of whom is without peer, both as diver beneath cold seas and gracious host.

We also wish to thank the following individuals for their help (listed in alphabetical order): Jane and Bob Altman, Beach Gardens Resort, Fred Calhoun, Cambridge Trust Company, Ralph Waldo Emerson Inn, Lou Fead, Janet Flannery, Roger Guerrette, Roger Hess (captain of *Charisma*), Jim Ingram (captain of *Westerly*), Dennis Judson, David Katzenmeyer, Ken Kivett, Coralee LaFresnaye, Brian Lyons, Shirley McLaren, Walt and Louis Musser, Ellen Oakley, Rene and Nate Rotman, Joyce Sizemore, Julio Stopnicki, TRW Sea Divers, Wayne Welch, Ramon Wenzel, and Jody Willoughby.

1/ THE COLD WATER WILDERNESS

Beneath Cold Waters

I HAD just decided to relax when the telephone rang. An insistent female voice identified itself as belonging to a reporter for one of the well-known national newsmagazines. The magazine was going to do a cover story on diving, and she had heard that I knew something about the subject. We talked about many aspects of diving, but most of her questions could have been reduced to the simple query: "Why dive?" I explained that I experience something special underwater, a way of seeing that is different, a freer way of moving—perhaps floating or gliding comes closer—and a new way of feeling. But my words seemed feeble in comparison with my experiences, so I resorted to an artifice I often use when trying to satisfy the unanswerable questions of "nonbelievers"—nondivers. "I can't really explain why I dive, but I could show you," I said. She brushed aside my offer of lessons and a ritual baptism. "Too busy," said she. And so my words, as inadequate as I felt they were, had to satisfy her curiosity.

She asked another question. I had better luck answering this one. "Why dive in New England?" This question rests on a false assumption: many nondivers assume that the only good diving is in warm tropical water, on a coral reef. The reporter's question was really quite general and could as easily have been Why dive in a California kelp forest?, Why dive in Puget Sound?, Why dive in Atlantic Canada?, or Why dive in British Columbia? In fact she could have asked me why anyone would dive along the hundreds of miles of coastline in the cold-temperate regions of the United States and Canada, waters in which millions of divers annually experience their unique sport. I answered with an analogy. We all know that a tropical jungle (even if we have never seen one) like a coral reef, is an exotic place filled with lush growth, brightly colored birds, insects whose wings have the glint of metal, and many other strange creatures. But those who have roamed a northern woods, a midwestern prairie wilderness, or a mountain meadow know that these places can be as interesting and offer as much as the jungle, though like their colors—earthy greens and browns—their pleasures are subtler than the jungle's bright hues.

I told her that there are many good places to dive and many kinds of diving, even in cold-temperate waters, where schools of brown-bodied cunners dart and play among granite boulders clothed with rockweed, where solitary lobsters lurk in holes and where hermit crabs scurry sidewise wearing the abandoned houses of dead snails (34). I do not think she believed that cold waters could be interesting. Nonetheless I have had many beautiful experiences and have seen much that is wonderful and exotic beneath cold waters. I have seen a richness of animal life and lushness of plant growth that would surprise anyone who has not shared my experience. I have seen a mob of small fish compete for the same morsel (37). I have marveled at a complete community of organisms confined in the space of a small tide pool that is flooded by the sea only at high tide (8). I have taken pleasure in the microcosm that lives beneath an overhanging rock: pink hydroids that look like elegant flowers, crumb-of-bread sponges, and white, brown, orange, and red tunicates (30). I have delighted in the coralline algae that look like bright spatters of pink paint on rock (11). I have watched barnacles underwater at flood tide with their opercula opened wide and their feeding appendages waving, changing the hard appearance they present when high and dry to one of soft and moving life. I have also dived on starry nights, and after my eyes had adjusted to the faint illumination underwater, I could see well enough to discern the gravelly bottom and make out an occasional shadow shape, perhaps a blackback flounder (3, 4) or a lobster wandering far from the hole in which it takes daytime refuge. I looked up and saw the canopy of stars. They seemed even more remote than usual and guttered with a cold fire. Underwater there were twinkling lights of another kind, the single-celled bioluminescent *Noctiluca scintillans*. I switched on my submersible lamp. In seconds, my field of vision contracted to the narrow cone of the yellow beam. I was half-blinded by reflections from the underwater snow of suspended particles. The *Noctiluca* had vanished. I killed the lamp. The bottom slowly rematerialized, and the living lights rekindled.

Some of my most beautiful experiences have been the simplest, the least dramatic. To look up and see the surface dimpled by rain or corrugated by a breeze, to have a three-inch fish attack my faceplate's shiny rim, or just to be still and slowly drift down, as if into a deeper and deeper sleep.

What Is Cold Water?

You can perform an interesting experiment: fill a basin with water and leave it overnight at room temperature. The following day, plunge your hand into the water. It feels strangely cool to the

2

touch. The water must have reached the same temperature as the air in the room, yet it feels cooler.

This phenomenon is no illusion. It is caused by the greater capacity of water to absorb heat from warm objects—your hand, for example—in comparison with air at the same temperature. This means that water at room temperature, 70°F (21°C), feels cold and water at 60°F (16°C) feels very cold. In fact, research gathered on shipwreck victims in World War II has shown that 60°F water can be lethal after just a few hours of immersion.

Thus, when we talk of cold water, we should not think in terms of air temperatures. However, our experience with air temperatures has shaped our understanding of the meaning of temperature. For example, we know that in air at 80°F (27°C) and of moderate humidity we can be comfortable wearing very light clothing; in air ten degrees lower we know that additional clothing is appropriate. Water is quite different. Water can feel uncomfortably cool at 80°F after a thirty-minute immersion. (Most indoor swimming pools are kept at approximately 80°F.) Water at 70°F is quite cold and is only tolerable for short periods or with the protection of a diver's wetsuit or drysuit, marvelous garments made of foam rubber that have opened up the beauty and pleasure of cold waters to millions of amateur undersea explorers.

Where on the coasts of the United States and Canada are sea surface temperatures at or below 70°F (21°C) for much of the year? On the entire Pacific Coast, from British Columbia down to San Diego, and on the Atlantic Coast as far south as the eastern shore of Cape Cod. In the coldest parts of the year, the surface temperatures of these regions fall as low as 50°F (10°C) in the Pacific and 28°F (−2°C) in the Atlantic.

What Makes Ocean Waters Cold?

The sea is a great engine, driven by the sun's heat and the spin of the earth. As the sun warms the atmosphere, winds are created. Great bands of wind prevail in the northern hemisphere: easterlies blow just to the north of the equator, westerlies begin north of a latitude of 40° (approximately the level of Philadelphia).

These great wind systems set water into motion in both the Atlantic and the Pacific. Because of the spin of the earth, these water movements are to the right of the wind. (To understand why this is so, imagine rolling a marble toward the center of a spinning phonograph record. The marble would be carried to the left by the clockwise rotation of the record. Since the Northern Hemisphere is spinning counterclockwise, the shift is to the right.) Thus, there is a general flow of waters into the central North Atlantic and central North Pacific. In fact, the water in the central parts of these two oceanic basins is slightly higher than the water at the edges. Some of this heaped-up water eventually returns to the perimeters of these two ocean basins; this outward flow is east and west (toward the continents) in both oceans. But this flow, too, is influenced by the spin of our planet and directed to the right, producing great clockwise currents. Along the Atlantic Coast of North America is found the Gulf Stream, a great clockwise current, a river in the sea. In the Pacific, the predominant Northern Hemisphere current is the Kuroshio (Japan current); it turns eastward and crosses the Pacific as the North Pacific Current, bringing temperate waters to the North

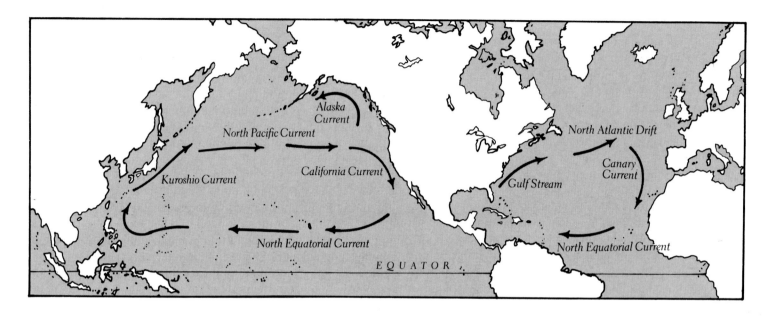

American coast at the level of Oregon and Washington. There it splits into a north-flowing Alaska current and a south-moving California current.

On the Atlantic coast of North America, the Gulf Stream sweeps tropical water to the north, creating warm-water conditions farther up the coast than might be expected. In the Pacific, the Alaska and California currents push cold-temperate waters as far north as Alaska and down the North American coast as far as the southern tip of California. The result of these diverging Pacific currents is that the waters from British Columbia down to San Diego are remarkably uniform in temperature, ranging between 50° and 70°F (10° to 20°C).

Vast numbers of temperature readings have been taken of the ocean surface from ships at sea in different months of the year. When these readings are plotted on an oceanic map, it is revealing to connect points of equal temperature with lines, termed isotherms. A glance at a map of maximum sea-water surface temperatures (usually reached in August in the Northern Hemisphere) shows the isotherm of 70°F to extend only as far south as Cape Cod on the Atlantic Coast, but in the Pacific during the same month, this cold-water isotherm reaches as far south as San Diego.

These, then, are the cold-water regions: the rocky shores of New England and of Canada's Maritime Provinces, California's vast and varied coast, and the rugged open coasts of the Pacific Northwest, but particularly the immense, protected marine waterways of Washington and British Columbia.

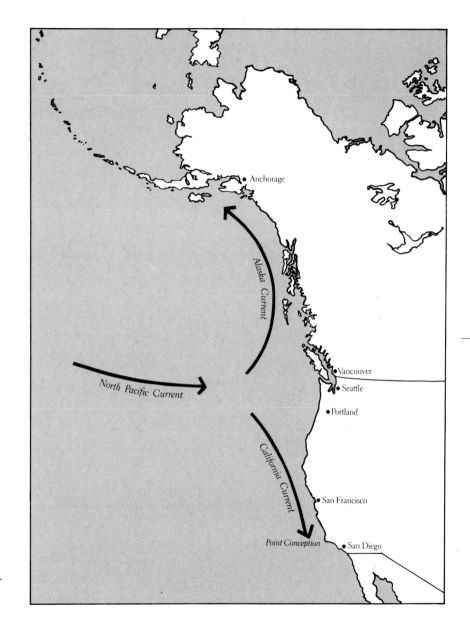

Figure 1.1

OPPOSITE: *Great clockwise water movements called gyres ring both the Pacific and Atlantic oceanic basins and greatly influence sea surface temperatures. Many other important currents flow in these oceans, but have been omitted from this diagram for clarity.*

Figure 1.2

The west-setting North Pacific Current bifurcates to form the Alaska and California currents, creating a fairly uniform cold-water region along much of the west coast of Canada and the United States.

5

THE COLD WATER WILDERNESS

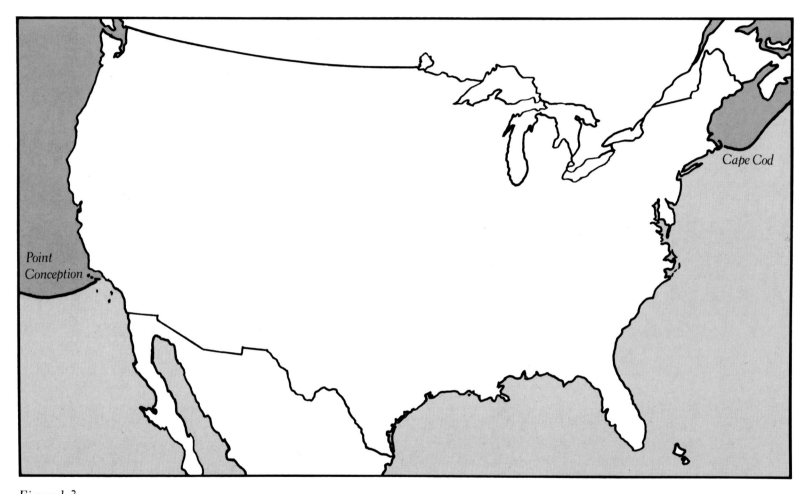

Point
Conception

Cape Cod

Figure 1.3

*Ocean surface temperatures (summer maximums shown) reveal
that cold waters of 70°F and below (darker area) are distributed
quite differently along the east and west coasts of North America,
reflecting the powerful influence of ocean currents.*

Characteristics of Cold-Water Marine Life

Some marine organisms can live in warm or cold water; these are called eurythermal plants and animals. However, the majority of marine organisms are stenothermal; that is, they tolerate or prefer a fairly narrow range of water temperatures. Most cold-water forms are restricted to cold waters. Thus, the cold-water environment is inhabited by a unique community of organisms, a community that cannot be observed in warmer waters. It is also characteristic of cold-water life that there are fewer different species of plants and animals here than elsewhere, but within those species there are greater numbers of individual organisms. Cold waters therefore can be rich in life.

Cold waters are generally less transparent than tropical waters for a variety of reasons, including the fact that many of the United States' most densely populated regions occur along its cold-water coasts. These developed areas result in the movement of great quantities of suspended matter into the sea from land, carried in fresh-water runoff and sewage effluent. In addition, runoff and sewage often contain nutrients that promote the growth of plant plankton (phytoplankton), free-floating, often microscopic plants, further reducing underwater visibility. The greatest water clarity tends to occur farthest from populous coasts.

There is another, subtler reason for the decreased transparency of cold waters: cold water promotes the growth of phytoplankton because it can hold a greater quantity of dissolved carbon dioxide, which is essential to photosynthesis. This capability is vitally enhanced by the mixing of deep nutrient-rich waters with the sunlit surface layers where nutrients are consumed by phytoplankton growth. Mixing may be accomplished by upwelling or by convection. Upwelling can be produced by a variety of conditions and is pronounced off the coast of California, where steady winds blowing from the northwest combine with the earth's rotation to move surface water out to sea. The displaced surface water is replaced by colder deep waters that rise up the steep continental shelves. Convection occurs when seasonal cooling in temperate climates increases the density of surface waters, causing them to sink. As in the case of upwelling, nutrient-laden deep waters rise to replace the displaced surface waters, a process called turnover. In both upwelling and turnover, the rising deep waters are well supplied with plant nutrients because of the continual rain of organic substances from the surface layers; these substances accumulate in the lightless, and hence plantless, deeps.

Convectional mixing is seasonally intense in temperate waters—our cold-water regions—because of the great annual variations in temperature that occur; temperatures vary little in the tropics (where they are uniformly high) or near the poles (where they are uniformly low). It is true that convectional mixing takes place constantly in polar waters, where the surface waters are usually colder than those beneath them, but the much lower quantity of available sunlight seasonally limits phytoplankton growth.

Increased phytoplankton growth results in an abundance of the animal plankton (zooplankton) that graze on them. This rich growth of plant and animal plankton greatly reduces the underwater visibility, especially during periods of explosive growth, called plankton blooms.

Because decreased transparency is a characteristic of

cold-temperate ocean waters, light does not penetrate as deeply in these waters as in warm seas, so plants of all kinds—as well as the animals sheltered or fed by them—cannot grow as deeply. The compensation depth—the greatest depth at which a green plant can collect sufficient sunlight to produce new tissue by means of photosynthesis—is shallower in cold waters. This means that we do not have to dive to great depths to enjoy the richness of cold-water life.

The Cold-Water Diver

All of us have spent by far the greater part of our lives breathing air from the atmosphere: walking, running, seeing, smelling, touching, all in the ocean of air into which we were born and within which our species has evolved. Every day we make hundreds of decisions based on our individual and species experience as air-breathing, terrestrial creatures. We decide how fast we can move without getting out of breath; what to wear when the air temperature is, for example, 70°F (21°C); what speed to drive in rain, fog, or at night; the precise position of an object we wish to reach out and grasp. All these choices, and many, many more, are seemingly trivial, but this fact is in itself a great marvel: our accumulated and catalogued experience enables us to do what we do so easily. A diver's objective is to achieve comfort and proficiency while moving through a fascinating but alien medium, and to do this he or she must overcome a number of barriers that grow out of the differences between water and air.

Of course, the most important way in which the water differs from the world above its surface is that water is unbreathable, at least for humans. In 1943, Jacques Cousteau and Émile Gagnan gave practical reality to an idea that had been around for a long time—taking air to breathe underwater. Today's self-contained underwater breathing apparatus (for which the acronym is *scuba*) is nearly identical in principle to that originally patented by Cousteau and Gagnan. However, the sleek, contemporary appearance of modern apparatus reflects refinement and sophistication as well as improvements in reliability, ease of use, durability, and cost. These developments have helped open up the underwater world to millions.

Cold is the next barrier that must be overcome. In all but tropical waters, divers cannot be comfortable for more than a few minutes without thermal protection. (Recall that divers, unlike swimmers, move through the water slowly when things are going well and therefore do not generate as much body heat as swimmers do from their increased muscular activity.) Therefore, second only in importance to the development of practical breathing apparatus was the development of diver's exposure suits: first the wetsuit of insulating foam rubber, and later the insulating drysuit. Although modern underwater breathing apparatus has opened the underwater world in general, exposure suits have been instrumental in opening the cold-water realm in particular. Divers equipped for comfort in cold waters are shown in the illustration.

Several additional factors make the underwater environment strikingly different from that of ordinary experience. Consider vision. Imagine swimming in a huge tank of water of the greatest possible purity; you would doubtless have a strong subjective sensation of extreme clarity. With the aid of a diver's face mask you would be able to see more than 300 feet

Figure 1.4

Clad for comfort, two divers prepare to explore the cold waters off a rocky coast. The diver on the right is dressed in a full wetsuit; *her companion wears a* drysuit. *Both garments are constructed of a synthetic rubber into which bubbles of insulating gas are blown. The wetsuit is form-fitting but not watertight; the small amount of water that does enter is quickly warmed to the temperature of the diver's skin and then causes no discomfort. The drysuit admits no water and is comfortable in the coldest ocean; it is sealed with*

the special airtight zipper visible on the diver's arm. Both divers are equipped with weightbelt, facemask, snorkel, and inflatable vest which gives buoyancy control underwater and flotation at the surface. The divers will also wear scuba, a knife, depth gauge, watch, compass, and fins.

(approximately 100 meters). Above water, however, this visibility is equivalent to that encountered in a moderate fog, a heavy rain, or a snowstorm. What seems excellent visibility underwater is poor visibility indeed by our usual standards. Thus, the underwater environment affects how we perceive reality and in quite a dramatic way.

Underwater, our vision is affected in at least two further ways. First, the diver wearing an air-filled face mask underwater perceives objects to be larger and closer than they are, and this requires, at first, some adjustment on the diver's part. This phenomenon is due to the fact that the air in the diver's mask acts as a lens. Second, water acts as a filter that selectively absorbs certain colors from the sunlight falling on the ocean surface, and this effect increases with depth. The visible spectrum ranges from red at one end through orange, yellow, green, and blue to violet at the other. Wavelengths of light just beyond those we can see are termed infrared (from the Latin *infra*, below) and ultraviolet (Latin *ultra*, beyond). Red light has less energy than violet and is therefore markedly absorbed in water, a far denser medium than air. Red disappears altogether at relatively shallow depths, even in pure water.

What, then, does a red fish look like when red light is attenuated and then extinguished? It may look orange, brown, gray, or black, depending how much red illumination has been lost. This is why underwater vistas appear first green, then blue,

THE COLD WATER WILDERNESS

and finally a deep indigo as a diver descends. However, any artificial light brought down, such as the underwater photographer's electronic flash or the diver's underwater lamp, will restore the lost colors. Therefore, a good underwater lamp is useful for revealing the true, vivid colors that do exist beneath cold waters. And for underwater photography below the first few feet, a good underwater flash is essential.

Today's amateur explorer beneath cold waters has the opportunity to experience a wilderness that is just an hour or two from dozens of large coastal cities in the United States and Canada. To find a place so relatively wild and exotic could, if above water, require traveling hundreds or thousands of miles. If the idea of exploring the world beneath cold waters intrigues you, start your exploration by seeking nationally certified training in diving, as from members of the National Association of Underwater Instructors. The pleasures of cold-water diving can turn to perils for the untrained or the unwary. High-quality diving instruction can be found in most United States and Canadian cities.

Even more challenging and rewarding than diving in the cold-water environment is diving and taking photographs. The underwater photographer distills, preserves, and shares the diver's unique experiences. Even though underwater photography was pioneered by Louis Boutan in 1893, it is still a relatively new art. The problems of photographing in and through water have yet to be solved entirely, and capturing photographs in cold waters presents a number of special challenges. The photographer's hands are protected from the cold with thick rubber mittens that greatly reduce dexterity. Then there is the problem of visibility; to take a clear photograph that is sharp and shows good detail, the rule is the closer the better. Thus, a wide-angle lens is standard for underwater photography, since it permits the photographer to encompass his subject fully when shooting from close range. But distortion is inherent in wide-angle lenses, and in many cases a certain amount of distortion has to be accepted. This problem occurs in the crystal-clear waters of a coral sea but is infinitely compounded in cold waters, where visibility may be less than fifteen feet (four-and-a-half meters) and discernible particles may be suspended in the water.

Lighting is another problem. Not only does the intensity of available light decrease with depth but, as we have noted, colors disappear selectively—reds, oranges, and yellows first—so that at 100 feet (30 meters) blues and greens predominate. Artificial light is needed to restore the brilliant scarlet of a sea urchin photographed at 60 feet (18 meters), or the lapis lazuli of a nudibranch at 100 feet (30 meters). Because suspended plankton and silt in the fertile temperate waters reflect the photographer's light, only certain lighting angles can be used successfully. If the photographer's subject is an animal in a crevice and the desired lighting angle is impossible, patience is required. Perhaps the animal will move to a better location, perhaps it will not.

Many of the animals in this environment see us as intruders, perhaps as predators; to photograph these animals in their habitual setting while engaged in their natural behavior often requires a slow careful approach, a certain amount of luck, and sometimes an offering of food. The subject must not feel threatened.

It is a great satisfaction to return from a dive, develop a roll of film and be rewarded with a photograph that captures the special beauty of some aspect of the cold-water environment. You are reminded once again that the pleasures of taking photographs beneath cold waters greatly exceed its frustrations.

2/ NEW ENGLAND WATERS

Rocks and Tide Pools

TYPICAL OF the northern New England coast, as well as much of the exposed coast of Canada's Atlantic Provinces, is a rugged shore of weathered granite. From Eastport, Maine, down to Cape Ann, Massachusetts, there is an almost unbroken stretch of rocky coast. However, in places—as at Plum Island, Massachusetts—there are long sandy beaches deposited by surf and shaped by along-shore currents. The cold-water region extends as far south as Rhode Island and the eastern shore of Cape Cod.

A piece of New England coast is often a tumble of rocks, as well as a profusion of broken angles, crevices, and rounded boulders. But the most striking feature of this shore is its tide pools. Some are flushed with clean sea water at every rise of the tide. Others are reached only by the highest tides, the spring tides, that occur about every two weeks.

A tide pool is an ocean in miniature. It may shelter an entire community of plants and animals, including rockweed, small crabs, and even fish. The careful observer may discover tiny isopods and amphipods (31), relatives of the shrimp. A great deal can be learned about marine life by patient observation of a tide pool and its denizens.

New England is famous for its tide pools. They are partly the result of geology: the native granite as well as the cataclysmic forces that have broken it to make flat terraces in which small basins catch the tide. But tide pools are also prevalent here

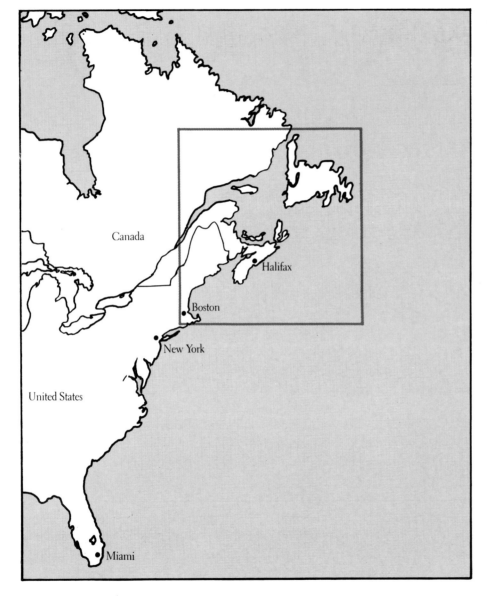

Figure 2.1

The cold-water diving region on the east coast of North America, bounded to the south by the 70°F isotherm, extends down as far as Cape Cod.

Numbers in parentheses refer to photographs.

BENEATH COLD SEAS

because of the great tidal range of the northeastern United States. The tidal bulge created by the attraction of the sun or moon, or both, on the ocean surface is exaggerated as it moves from midocean to the shallow continental shelves; the gentler the slope of the shelves, the greater this effect. The continental shelf off New England is especially broad as well as gentle in slope.

In Massachusetts, the tidemarks on the shore—the high-water or low-water limit of the tides—may be from eight to twelve feet apart. Further north, in Maine, the tide may rise and fall a full thirty feet, as in Passamaquoddy Bay, due to the very gradual slope of the bottom and the funneling effect of the bay. In any case, the great reach of the New England tides allows a broad stretch of rocks to be flooded at high tide and to be exposed at low tide. Any seawater caught in small pools will remain isolated from the sea until the next flooding of the tide.

A rocky coast in New England may seem a confusing picture to the casual observer; some order is brought to this seeming chaos by a collection of forces that together create vertical zonation. Dramatic changes occur in the distribution of plants and animals from the highest reach of high tide down to the lowest extent of low water. Those hardy organisms that live at the top of this zone must resist extreme environmental change, for when the tide falls they are exposed to the drying effects of the atmosphere and the sun, and are no longer shielded from the sun's intense and often lethal ultraviolet rays. (The harmful effects of these rays are screened out by just a few feet of water.) A little lower in the intertidal zone (the zone between tidemarks), the periods of exposure to the atmosphere are shorter; thus there is less drying, briefer exposures to ultraviolet radiation, and smaller fluctuations in temperature. Even lower in the intertidal zone, where the bottom is submerged more than half the time,

still gentler changes occur. However, it is not until well below the zone of the lowest of low tides that environmental conditions can be said to be relatively constant. The intertidal zone provides a range of environmental conditions—a range of niches, distinct environments in which organisms can live and to which they must adapt. Organisms assort themselves from top to bottom in the intertidal zone according to their resistance to environmental change.

The rocks at the top of the intertidal region are blackened by a simple but very hardy single-celled organisms, a species of blue-green bacteria, cyanobacteria, that until quite recently were thought to be unicellular algae. Below this black zone are the barnacle zone and the periwinkle zone. There is no sharp demarcation between the barnacle and periwinkle zones; the animals that dominate these zones (and after which they are named) often mingle. Barnacles are crustaceans, cousins to the crab and the lobster. Unlike their mobile relatives, the barnacles cement their calcareous houses to rock. When the tide moves out, each barnacle closes the opening within its shell with movable plates called opercula, thus retaining a drop of life-sustaining moisture. At flood tide, when the barnacles are covered with water, they open their opercula and extend their feeding arms (44) to feed on plankton. The multicolored periwinkle, an intertidal snail, likewise traps water in its shell and seals itself tightly to the rock when the tide is out to protect itself against lethal desiccation.

Lower, below and seaward of the zone of barnacles and periwinkles, lies the rockweed zone. The term *rockweed* includes several species of fleshy algae that have a striking characteristic in common: all possess gas-filled sacs. When the tide is out, these fleshy plants collapse on each other and form a thick mat that holds moisture. At full flood the algae are buoyed up by their gas-filled floats; they stand up and separate from their neighbors, allowing space for water and sunlight. These algae have uniquely adapted themselves to life in the intertidal region. Other organisms find shelter within the rockweed bed; small crabs are adapted to move among and across these algae and are well matched in coloration to their algal hosts. Minute amphipods and isopods, as well as the exotic nudibranchs, shell-less mollusks (16, 17), may also be seen here.

Irish moss, *Chondrus crispus*, also an alga, carpets the rocks just below the rockweed zone. In the deeper part of the Irish moss zone, the Irish moss is rust colored, even maroon; in shallower water it is bleached yellow or white. At high tide a wonderful fish, the sea raven, *Hemitripterus americanus*, conceals itself in the Irish moss; it is maroon, rust, brown, or yellow and has fleshy, plantlike appendages. It is easily overlooked when it rests motionless on the algae (1, 2).

Deeper lies a region that is never exposed to the atmosphere: the sublittoral zone. It is also called the kelp zone in recognition of its dominant organism. Kelp, the largest of the algae, begins to grow at the low tidemark and can be found all the way down to the compensation depth. Kelp growing in the uppermost part of the sublittoral zone are occasionally exposed to air at the extreme low tides, which occur four times a year. The kelp in New England occur in a variety of forms and harbor a diverse collection of other organisms, great and small.

The rocky New England bottom provides attachment and shelter for many plants and animals. Mobile forms such as the lobster, *Homarus americanus*, whose defense is its armor and powerful claws, inhabit crevices and fissures in the rock. The lobster is best protected with its back in a hole and its eyes and antennae alert; it confronts the world with its pincers (15). Boulders are spattered with pink coralline algae. The crumb-of-bread sponge can be seen in sheltered spots. Occasional sea anemones sprout from rock, like wildflowers on a mountainside (20). Crabs scuttle boldly over the granite or cower timorously in any available recess; rock crabs, Jonah crabs, green crabs, and the ubiquitous hermits can all be found here (13, 14, 34). Clumps of horse mussels, *Modiolis modiolis*, hold onto the hard bottom with their byssus, tufts of sticky filaments (30).

Deeper in the sublittoral zone, gregarious green sea urchins grow in beds on a rock bottom (11); when these beds are large, the kelp is sparse—and for good reason. The sea urchins appear passive, but are in fact voracious consumers of kelp. These bottoms are also studded with small and medium-sized sea stars (starfish), which also look docile and sluggish enough. Turn one over and you will see a central mouth and many small tubular appendages, called tube feet. Each tube foot is tipped with a suction disk. The tube feet are connected to the sea star's water vascular system, a living hydraulic apparatus with which the animal can create suction in any or all of its tube feet. The organism uses its tube feet to cling to the bottom, move over the rocks, or grasp prey. The sea star's simple nervous system precisely controls each tube foot, as if it were a single individual in a marching army and coordinates their separate movements to take the whole in the desired direction.

Sea stars and sea urchins are closely related; both are echinoderms (from the Greek for spiny-skinned). The underside of a sea urchin also contains a central mouth and an array of tube feet. Imagine a sea star without arms and with the hard bumps in its skin grown to long spines: you have a sea urchin. A sea urchin feeds on kelp, whereas a sea star consumes the flesh of living

14

mussels as well as carrion and other organic debris. How can a sea star open a mussel? It creeps along the bottom until it is astride its prey, then it grips the shells in a fatal embrace, creating a small but inexorable suction with its tube feet (8). Eventually the doomed mollusk can resist no longer and relaxes its powerful adductor muscle. The sea star then extrudes its stomach through its own mouth and into the narrow gap between the mussel's shells. The echinoderm's gastric glands secrete juices that begin to digest the mussel in its own house; the sea star then absorbs this partly digested soup. The slow violence of this struggle will pass unnoticed to the casual diver, but patient observation will be rewarded.

Small fish also inhabit the sublittoral zone. Hoards of brown cunners, *Tautogolabrus adspersus*, related to the colorful wrasses of tropical waters, play among the kelp and bottom rocks. The occasional flash of a silversided cod can be seen in the periphery of the diver's vision. Only small ones live in these shallow waters, and they are quick and cautious. A really fortunate diver will see the elusive thick-lipped tautaug, also a wrasse.

I Know a Place

The rocky Atlantic shore with its rich, zoned array of marine life offers marvelous experiences for the cold-water diver. I have had many such experiences, and have chosen one as an example of the best you can expect.

Three of us stood on a smooth rock that was just awash. The retreating waters of the last wave pulled at our swim fins. Every square inch of our skins was covered with the thick foam rubber of our wetsuits. I shifted my stance to ease my scuba, an uncomfortable burden on my back. The water-worn granite was slippery, especially where patches of brown rockweed clung to it. We did not risk walking. Instead, we waited as the water came to us. It rose, then we fell trustingly onto the full breast of the new wave and rode it out. We breathed through our mouthpieces as we swam down through the troubled surf zone. The water was filled with bubbles from the breaking waves.

We glided into a deeper zone. Here the water was clearer, and the rhythmic waving of kelp thalli was the only visible effect of the turmoil above. When we reached a depth of thirty feet, we turned to the right and leveled off. I kept my eye on my depth gauge, a bulky instrument strapped to my wrist and looking like a fat wristwatch with only one hand. We followed a line of constant depth, an underwater course that kept us roughly parallel to the irregular shoreline above. Soon we discovered a narrow inlet, entered it, and found ourselves in a tiny cove—more like a cave without a roof. I swam deeper into the slit; the others hung back and waited. (I have always derived an intense, even selfish, satisfaction from the fact that I was alone in this place.) I lay on my back in the deep, narrow space and looked up. The granite walls focused the incoming swells. I watched as the sea broke. This spectacle is always impressive when seen from above, but when observed from my vantage it seemed even more awesome.

And there was silence. The energy of the crashing water seemed greater in silence, like blinding lightning without thunder, or a tree-bending storm with no roar of wind. In a magnificent show of power, each wave rose and spent itself against the rock. The water whitened as bubbles filled it, then

NEW ENGLAND WATERS

cleared as they dissolved. The next wave humped up, then another and another. On the bottom, a gentler rhythm of advance and retreat swayed a grove of rockweed, like wheat in a breeze. A few small fish kept their stations, alternately heading shoreward and seaward, but always into the breeze.

New England Sand Bottoms

Sandy beaches look very much underwater as they do above: a flat expanse of gently sloping sand. At first view, this kind of shallow, soft bottom is a desert. Little of the richness that clothes the rocky bottom is evident here, yet a closer look reveals some interesting things. For example, a diver swimming across this bottom may observe what appear to be open mouths embedded in the sand. These are actually the siphons of the slender razor clams, whose shells and bodies are buried deep in the sand. This clam's ability to burrow is prodigious; it is nearly impossible to dig one out of the bottom. Dig after one and it retreats deeper into the sand. I have never been able to capture one. The slightest pressure, even the wake of a fin flick, will cause the sensitive tissue of the siphon to react, and the clam will pull it down through the sand to escape a potential predator. Once, when I was still trying to dig out razor clams and was not content to see them only in the form of dead shells littering the bottom, I thrust my diver's knife into the san beside a siphon, hoping that a quick prying movement would lever the clam out of the sand. Carelessly, I cut off the tip of the siphon by mistake. I touched it lightly; it contracted just as it had when still attached to the clam. It seemed alive; the escape reflex is inherent in the tissue and

does not depend on a central command from the bundle of nerve cells that serve as the clam's brain.

A sand bottom presents a serious problem to animals that inhabit it: the flat, unrelieved surface makes them easy pickings for predators. Any animal that ventures out across this unprotected plain must have some trick, some way to elude its enemies or to conceal itself from its prey. A common ploy is to be flat like the sand and to lie still upon it. This works for sand dollars (10), flounders, skates (6), and the marvelous goosefish, which deserves a section all to itself.

Goosefish I Have Known

People often ask me if I encounter sharks when I dive. They seem disappointed when I tell them that we rarely see them in New England waters. Fishermen may encounter big blues far offshore, but in the inshore waters in which we dive I have only seen the spiny dogfish, a small and harmless bottom-dwelling shark. There just are no big predators in New England inshore waters. You must have heard the story—we all have—of the great white shark, *Carcharinus carcharias*, the savage killer that ranges worldwide. He could, so the story goes, attack anywhere and anytime, without warning. The statistics, however, show little cause for alarm. The chance of you or I ever being attacked by a shark is very slim indeed. In New England this possibility is almost nonexistent.

What do we have to offer the thrill seeker who looks for danger underwater in New England? I suggest the dreaded goosefish.

The goosefish, *Lophius americanus* (5), is a shallow-water relative of the deep-water anglerfish, a grotesque inhabitant of the lightless deeps. The anglerfish sports a luminous organ at the end of a stalk protruding from the top of its head. The stalk and its lamp are deployed as fishing rod and bait: the anglerfish dangles the luminous bait in front of its mouth, which it holds agape. Any hapless fish that goes after the proffered morsel is likely to be swallowed whole.

Food is scarce in the deep sea, and few fish of any kind are to be found. Not only must the anglerfish be able to attract its supper, it must be able to eat whatever shows up, even fish bigger than itself. Many of the various kinds of deep-water anglerfish have adaptations that allow them to open their jaws wider than their own bodies. Many are equipped with roomy mouth bags so that when they open their mouths to the fullest, they can engulf a fish larger than themselves.

The goosefish of the New England sand bottoms shares with its deep-living cousins the ability to swallow big prey. It has a pair of jaws that function like the kind of pocketbook that has two semicircular hoops hinged together. When the pocketbook is shut, the snout of the goosefish is flat and thin, but when it opens wide there is a large circular maw, as high as the fish is wide. The skin attached to the jaws is loose and flexible, so the gaping jaw becomes a bag capacious enough to hold a big fish.

The goosefish also has a rod and bait. I have been privileged to see it raise its baited fishing rod from a retracted position along the top of its snout and dangle it just ahead of its own closed jaws. This happened when I had removed my snorkel from its usual position—attached to the head strap of my face mask—and offered its brightly colored tip to the goosefish. Its sluggish brain must have told it that here was something to eat. It erected the

stalk on its snout, canted it forward, and began rhythmically to wave the little fleshy tassle that fluttered from the end of it. I knew enough about goosefish not to reach out and tickle it with my gloved fingers. (I had seen the sharp teeth on a set of dried goosefish jaws that one of my friends possessed. He showed me how large the jaws were when opened by slipping the toothy circle over his head and wearing it around his neck, a grisly necklace.) I poked the fish's thick lips with my snorkel a couple of times. No response. The goosefish is generally a lazy fish, but this one, a five-footer, finally did open his jaws. Perhaps it was just a yawn. I peered into the great mouth. Here was a beast fully the equal of the great white shark.

I became a little bolder and decided to take a few more liberties. I swam around behind the fish and grabbed it by the very tip of the tail. It struggled and I felt its considerable strength. It was not limber enough to bite its own tail, so it could not reach me, but I had quite a job to hold onto it. I hoped it would get really angry; perhaps it would swim, and I wanted to see how it propelled itself and what it looked like underneath. Up to now I had only seen the wrinkled and mottled skin of its back, its two bulbous eyes, and the fins at its sides and tail. There was no sign of gills. I shook it gently and released it. This time it was angry. It waved its tail in broad, lazy sweeps and glided about six inches above the bottom. I still saw no gills, but I did see the vents through which it exhaled water that had passed over its gills. These vents looked like the exhaust cowling of a large jet. You might laugh at me for saying so, but the goosefish had a certain rough dignity and rustic grace that appealed to me. It seemed beautifully adapted to its flat, sandy environment. It was absolutely smooth underneath, perfect for lying on the bottom. Its pebbly skin was good camouflage, and its jaws were a

17

marvelous piece of machinery.

In my corner of New England, one of the most popular places to dive is in a well-known cove. Here, the favorite underwater route is against the sheer rock wall that forms the western boundary of the cove. We usually enter the water at a small rocky beach and then swim on a certain compass bearing until we reach the base of the rock wall, about twenty feet (six meters) underwater. Then we turn to the northeast and follow the base of the wall out toward the mouth of the cove. If we swim all the way out of the cove, into the open sea, we can reach a depth of fifty-five or sixty feet (seventeen or eighteen meters), depending on the state of the tide. I have made this dive many times. There is always much to see, since the rock wall serves as a firm base for the attachment of a variety of plants and sessile animals, which in turn provide food and shelter for fishes and small crustaceans. On a day when the water is calm and the underwater visibility is good, this can be a spectacular dive.

I have made this dive so many times that I know many of the underwater landmarks by heart. I can recognize a certain boulder, a particular rock slab, even a concrete block that has rested in the same place on the bottom for several years. When I see one of these familiar sights I know where I am. At one point along the wall there is an overhanging rock with a broad but shallow cave underneath. For many seasons, this was the home of the resident goosefish. I could expect to find it here; I might stop to play with it. Perhaps I would hold its tail and try to make it angry, or break off a cluster of blue mussels adhering to a nearby rock and see if it would try to swallow them when offered. Once I got it to swallow a bunch of five or six, which it spat out almost at once. After paying my respects to the goosefish, I moved on. The goosefish became a mascot for all who frequented this particular place. We would tease it, poke it, and holds its tail—but always gently and with respect. It was the good spirit of the cove.

I will never forget how I felt one day when I drove up to the small dirt parking area adjacent to the cove to spend a day diving. I saw a goosefish splayed on the roof of a car: a spear pierced its head just behind the right eye. The animal was some man's trophy. I swallowed hard a couple of times. I walked over to the car and saw a young boy and his father just out of the water, each in a wetsuit and each with a spear gun. I tried to make friendly conversation. Casually, I asked them about the goosefish. The boy said, "Well, he attacked me. I had to spear him, and he doesn't do anybody any good anyway." I patiently tried to explain to the boy and his father that we had come here many times and that the goosefish had never bothered anyone, and that it was a shame to kill it; and if they could not eat it, it would be a waste. I was fighting back my anger; I knew if I allowed my fury to show in my speech or manner, I would lose any chance I might have to persuade the boy and his father to give up their catch.

I looked at the goosefish. It was impossible to know whether it was alive or dead. I tried to convince the pair that they should remove the spear and restore the injured beast to its element. I hoped it would be able to swim away. I looked at it again; its tail twitched. There was hope. I finally talked them into it. The three of us lifted the goosefish carefully down, cradling it underneath. I unscrewed the spear point and gently withdrew the shaft. We set it down in shallow water. I grasped its tail, but it showed no sign of being alive. I moved it back and forth in the water for a few minutes, trying to force water through its gills. After a time, I felt a muscle twitch in its tail. I must have continued my efforts at resuscitation for half an hour until I was certain it was alive. Finally, it started to make feeble swimming movements. I was never so happy as when it was able to swim away. I returned to the parking strip, suited up, and my companions and I entered the water and started our dive. I looked for the goosefish in its usual lair. It was gone. I have never seen it since.

Other Inhabitants of Sand Bottoms

Certain inhabitants of sand bottoms rely on armor, instead of disguise, to protect themselves against their enemies. These include hermit crabs, other crabs, the moon snail, *Lunatia heros* (25), and the lobster. A third category of organisms are neither flat nor armored but depend on near invisibility; this includes the almost-transparent sand shrimp.

The sand dollar, *Echinarachnius parma*, with a flat body, relies on an additional factor to insure its survival: it has very little edible flesh. It has few enemies (the flounder is one) because it can provide a predator with little nourishment. The sand dollar is an echinoderm, like the sea star and the sea urchin. It, too, has radial symmetry and various other features common to all echinoderms.

Eelgrass, *Zostera marina*, is an interesting inhabitant of some sand bottoms. A true land plant, it produces seeds, unlike algae, which reproduce asexually or through spores. (True seed-bearing plants which live underwater are called sea grasses; only multicellular marine algae are properly referred to as seaweeds.) It is thought that eelgrass represents the adaptation of a terrestrial plant to the coastal marine environment. Eelgrass is found totally submerged, in some places as deep as twenty or thirty feet (six to nine meters). More commonly, however, it is found in shallower waters, ten to fifteen feet (three to four-and-a-half meters) deep. Beds of this sea grass render the flat sand bottom conducive to the survival and prosperity of a community of organisms specifically suited to live among its grasslike blades. A striking example is the pipefish, *Syngnathus fuscus*, whose long, slender body can easily be mistaken for a blade of eelgrass. The pipefish assumes a vertical position among the sea-grass blades. Spider crabs also live within the shelter of the eelgrass bed.

Another inhabitant of the sandy bottom, the horseshoe crab, *Limulus polyphemus* (28), is one of the most ancient animals. In spite of its name, it is not a true crab, but a member of the subphylum Chelicerata, which also includes spiders and scorpions. The horseshoe crab's jointed carapace looks like a medieval helmet. If you turn one of these animals onto its back you will see that it has five pairs of walking legs, as do true crabs. (True crabs are decapods, meaning ten-legged.) Solitary horseshoe crabs move slowly across the sand. Often, however, they will be seen moving in coupled pairs, or even trios, a smaller individual apparently hanging on to the tail (telson) of a large specimen. Such a pair will maintain their close connection even when moving over obstacles on the otherwise flat bottom. I have even seen a ménage à trois—three horseshoe crabs connubially attached (29, 40, 41). I once turned a joined pair onto their backs, perhaps an inexcusable invasion of privacy, to see how they accomplished their connection. I noted that the forward appendages of the follower crab tightly embraced the telson of the leader. The smaller individual is the male, who climbs aboard the female from astern and maintains its hold with its first pair of

19

walking legs.

On certain sand bottoms I have encountered the blue-eyed scallop, which takes its name from the fringe of azure eyes visible in its mantle when the animal's shells are held just agape. These light-sensitive organs are able to detect movement as well as light and shadow; how well they are able to distinguish shapes is unknown. These are swimming mollusks; by sharply clapping their shells together they are able to propel themselves backwards to elude a predator. Their most feared enemy is the sea star. Scallops and other swimming mollusks have been observed to swim frantically to escape when their mantle is touched by a sea star's tube foot. I have encountered individual scallops on the sand and have seen them swim. But apparently scallops do not always live as solitary individuals. Friends have told me they have seen large beds of these deliciously edible animals, but politely decline to reveal their location and I have never happened onto one myself. Yet occasionally after a storm I have seen a litter of dead scallop shells on a beach, indirect confirmation that beds must exist. Someday perhaps I will find mine.

When closely observed, the scallop is a beautiful animal. Its just-parted shells show wreaths of purplish eyes, embedded in its creamy mantle. The shell itself has a certain satisfying symmetry that makes it the epitome of all shells.

The flounder is also a fascinating beast. It looks as if it had a losing encounter with a steamroller. It appears to lie on its belly, with its broad back exposed. However, if you look at it carefully you will see that it is lying on its side. Its wide tail rests flat, not perpendicular to the bottom as you would expect if the flounder were flattened top to bottom. Only a single gill slit is visible, and the mouth is curved into a half grin. Between the half mouth and the solitary gill, where you would expect to find a single eye, you

find two; the eye on the underside has migrated round the fish's head to join its counterpart on top (4). The flounder is not born this way: for several weeks after it hatches it swims about upright with its eyes on opposite sides of its head. As the larva grows older, it becomes progressively flattened from side to side and begins to list. Eventually, it leans fully over onto one side. The lower eye has already begun its upward journey.

Being left-handed myself, I began to wonder if flounders always favored the same side and were thus all right-sided or all left-sided. I have heard that most cold-water flounders are right-sided: both eyes are on the right side of their heads. On the other hand, tropical species are generally left-sided. However, left-eyed species do occur in cold waters, and a group of North Pacific species is named for this characteristic—left-eye flounders; this group includes the California halibut and the Pacific sanddab. But in my experience, the California halibut seems to come both left-eyed and right-eyed.

The flounder's other gill remains on the fish's underside, hidden from view. The flounder turns this arrangement to good account. The greatest benefit of the flounder's flatness is concealment on the bottom, but a visibly moving gill cover would betray it. Therefore, instead of breathing through its visible gill, it arches its body slightly to create a hollow along its underside. It exhales into this tunnel through its lower gill. When disguise is futile and flight is best, the recumbent flounder sharply closes its lower gill, expelling a jet of water which lifts it off the bottom and starts it forward.

In New England there are several kinds of flounder: the blackback flounder (or winter flounder), *Pseudopleuronectes americanus*; the yellowtail flounder; and the windowpane flounder. The blackback is the species most often eaten. The

20

windowpane flounder is so thin as to be translucent; it has little edible flesh and is of no commercial importance.

The appearance of a New England shallow sand bottom undergoes a striking transformation at night. Animals that spend their days hiding in crevices emerge under the protective cloak of darkness. Thus, the lobster roams abroad at night; and local laws generally forbid nocturnal lobster catching, because it is too easy and therefore too destructive to the lobster population. At night certain fish move from deeper water onto the shallow sand and seem to fall asleep. Flounders become sluggish and are easily captured, but fair play does not permit it. Bioluminescent microorganisms twinkle in the dark water. A night dive on a New England sand bottom can be a unique experience.

Night Dive in New England

It had been a warm fall day. The sun was about to slip behind the houses facing the beach. My students and I had planned to be completely dressed by sunset. We would then wait for ten or fifteen minutes, as twilight deepened into dusk, to permit our eyes to make their slow adjustment to night vision.

We were ready. We walked down the beach, fins in hand. There, we talked quietly together and checked our underwater lamps. Each of us carefully avoided shining his beam into another's eyes, since even a brief flash of bright light can wash away any dark adaptation that has already taken place.

Twenty-five minutes after sunset we donned our fins and walked backwards, slowly, into waist depth in the flat-calm water. Venus and the moon shared the sky. It was too early for the stars, but the sky was clear enough. We made a final check of our equipment and briefly rehearsed our plan. Then we lay down in the water and began to swim out and down. As agreed, we stopped at a depth of about ten feet (three meters), where we knelt together on the bottom and waited. Slowly, I began to see the sand; I recognized bottom features that I knew well from many daylight dives at this same spot. (It is a cardinal rule of night diving to dive only in places where you have dived, and learned to know well, by day.) I looked into the faceplates of my companions. I could not see their eyes, only black glass and an occasional glint of watery moonlight. I looked at each one and made the OK sign close to his faceplate; each responded with an OK. All was well.

I began our slow procession. I followed the tumbled line of rocks at my left, the base of a seawall. This is a fairly straight line, well demarcated by the rocks at the left and flat sand on the right. We followed this course until the sand gave way to gravel. I knew that this change occurred at a depth of about twenty feet (six meters). My direction-finding sense was automatic now; I was swimming a course I had followed many times in daylight, and I recognized some of the larger and more unusually shaped boulders. I knew by the nature of the bottom—sand or gravel or boulders—where and how deep we were. I was confident as I led the group. We stopped again. I decided this would be a good time to show my companions a surprise. I motioned for them to join me. We could see each other as silhouettes with no difficulty. When I knew I had their attention, I snapped my gloved fingers. Sparks flew from my fingertips. Then I moved my hand vigorously as if I were conducting an orchestra; a shower of stardust followed in my hand's wake, the glow from small single-celled organisms that give off a firefly light when disturbed. My companions soon snapped their fingers too, and moved their arms and hands to create their own luminous clouds. We played this game until I signaled that we should move on. One of the divers did not respond. I had to tap gently on his faceplate and beckon to him to come.

I saw a shadowy but familiar shape on the gravel bottom. I slowed my breathing and moved stealthily, almost gliding down onto the creature. I reached out cautiously and grasped the lobster, just behind the appendages that bear its claws. A lobster can be held safely this way (39); it cannot reach its captor's hands with its claws. I turned around and presented my catch to my companions, tail first. Each of them, one after the other, carefully held the lobster from behind and then passed it along to his neighbor. I knew from having done this in daylight and seeing the faces around me that this was a thrilling experience for a diver who had never done it before, and yet to do it at night is even more exciting. I vicariously partook of their new pleasure. When the last of them had handed the lobster back to me I held it up for a final time and then released it in midwater. It floated slowly down; it knew it was free. It folded its pincers together in front of its head, beat the water with a quick stroke of its flat tail, and shot backwards. Again, I could imagine that the others were amazed.

We swam on. I flicked the switch on my underwater lamp and its yellow beam changed the whole character of our underwater experience. Before, the light had been blue and cold, like pale moonlight; in fact, it had been moonlight and starlight. Now the light was harsh and yellow, and particles suspended in the water sent back strong reflections, like the light of headlights bouncing back from snowflakes in a blizzard. I could only see objects within the yellow cone of my lamp, so when I shined it on the bottom my vision was limited to a very small circle of light. I lost my sense of the bottom. I also lost the feeling of visual calm that had prevailed before. The garish light was like loud noise. The others responded and flicked on their lamps and there was a cacophony of yellow, blinding beams. I switched my lamp off and one by one the others did too. I have always found night diving more pleasant without lamps than with them, though lamps are essential items of safety equipment, and we always take them with us.

I wanted to check my depth. Though my depth gauge was equipped with large phosphorescent numerals, they were too faint to read. I employed a trick: I placed the lens of my flashlight directly on the face of the gauge and switched the lamp on. After several seconds I switched the lamp off and uncovered the gauge. Its numerals glowed brilliantly. This brightness endured a minute or two, long enough for me to read my gauge and to let everyone else in the group read it too. They all used the trick, which works on any luminous dial. Soon I looked at the dark human shapes around me and saw the bright dials of compasses, depth gauges, and underwater pressure gauges glowing with a greenish light.

I had warned the novice night divers with me that we must stay close together, much closer than in daylight. Early in the dive they followed my advice too well—they bumped into each other and into me, it seemed, with every fin stroke. Now, however, I could see they were more relaxed. They swam at a comfortable distance from each other and from me, yet we did not become separated. They were beginning to enjoy the night dive on this sand bottom.

It is more than just an adventure to dive at night. It is an experience that makes us aware of a fundamental feature of our world: simple space. I have always thought of diving as a spatial experience, opening up the freedom to move in three dimensions, and diving at night brings this point home even more clearly. At night I see the bottom, the lobsters, and, occasionally, a sleeping fish that may tolerate touching for a second or two. A solitary skate, startled by my lamp, may fly off into the darkness (6). But the overwhelming impression is of liquid space, of being immersed in a fluid medium whose tangible qualities are reinforced by the almost solid darkness.

I have long noticed that when diving at night I seem to consume less compressed air than by day. This may be an illusion. I have not done careful checking, but it seems we can stay underwater longer and swim further at night than under the same conditions by day. If this is so, there is a good reason for it. The human body's daily rhythm of activity follows a twenty-hour cycle, called circadian rhythm; when we dive at a time when ordinarily we would be sleeping, our need for oxygen ebbs and we stretch out the air in our scuba cylinders.

Now my students seemed to have the hang of it. They were swimming along with their lamps off, enjoying the liquid darkness. Their motion seemed smooth; they stayed with me yet we did not collide. We had gone out far enough across the gravel, and I knew we should make a great loop to return to where the gravel meets the sand and then retrace our earlier course to shore. I found a friendly boulder, shaped like the prow of an upended rowboat, that I knew as my landmark, the outer limit of our dive. I began our great circle; I am sure the students had no notion we were turning. I checked our heading with my compass.

We ran off the gravel onto the sand and found the line of tumbled rocks from the seawall. I had navigated this stretch too many times before to be more than mildly pleased by my accuracy. I did not rely solely on my compass or landmarks; I had a feel for the bottom here and knew it well. I would have been shocked if we had not come off the gravel and onto the sand near the point where the seawall stood. We followed the wall, this time keeping it to the right as we headed shoreward. I turned on my back, kept on swimming and looked up. I saw the faint twinkle of stars through the water. As we moved forward the bottom slowly rose up. I glanced up again and could begin to see the fine texture of the surface. We were very shallow. I skirted a boulder and stood up in chest-deep water. I stood quietly waiting for the others to imitate me. Soon they were all standing around me. They let their mouthpieces fall out of their mouths and brushed their masks onto their foreheads. They talked excitedly about how beautiful it had been, about this fish and that lobster and the bioluminescent *Noctiluca scintillans*, though they did not know the scientific name. It had been their "best dive."

Deep Bottoms

Modern techniques and equipment permit experienced amateur divers to explore the underwater environment safely to a depth of approximately one hundred feet. Although this is not deep in the vocabulary of the oceanographer, it approaches the technological and physiological limits for recreational scuba diving. In any case, a depth of one hundred feet certainly impresses the diver who experiences it—especially in northern waters—with a strong subjective feeling of being very deep indeed.

This is a twilight world. Even at eighty feet, the turbid New England waters permit only the feeblest rays of sunlight to reach the bottom. (In tropical water, by contrast, the same degree of light attenuation may not occur until depths of several hundred feet.) Swimming through a rubble of cyclopean boulders in this perpetual dusk is a magnificent and eerie experience, like entering a darkened cathedral whose nave, choir, and crossing are lit only by moonlight passing through stained glass. In fact, one of my favorite deep-bottom diving sites is named Cathedral Rocks. The boulders loom ahead, and as I pass them I have the certain knowledge that they are enchanted and take on a life of their own when they are at my back.

A deep bottom often consists of fine gray mud and is a featureless plain, except where hard surfaces permit the attachment of plants and sessile animals. Hard substrates can be boulders and rock outcrops, or such man-made objects as breakwaters of quarried granite or steel shipwrecks.

Plants are scarce at these depths, but animals are abundant and are often larger than similar forms occurring in the shallower and sunnier regions. An occasional grandfather of a lobster can be seen backed into its hole. Hanging gardens of good-sized sea anemones may clothe an overhanging rock surface; some may even grow upside down.

In some places on the New England coast a diver can reach deep bottoms with a moderate swim from shore, but it is on deep bottoms far offshore that a diver can begin to find organisms characteristic of the open sea, little influenced by the transitional environments of the shore (where water meets land) or the surface (where ocean meets air). These offshore deep bottoms are accessible to divers only by boat.

In the tenebrous light of offshore deep waters lives a variety of organisms rarely seen in shallower waters or in deep waters closer to the coast. Various deep-dwelling hydroids (related to sea anemones) live here. Their colors and flowerlike forms delight the eye. It is here that whole colonies of sea anemones are found, usually on overhanging rock surfaces that face the open sea. I recall gliding down along the base of a deep submarine cliff, examining the drab bottom beneath me, when a vague impulse prompted me to look up at the rock jutting out overhead. What I saw took my breath away: hundreds of anemones carpeted the rock. Their delicate tentacles were tinted brown, yellow, white, and orange. (Anemones are common in shallow depths too, but rarely occur in the size and profusion found here.) The dominant New England anemone is *Metridium dianthus* (18, 20, 21). Other species are also seen, usually as solitary individuals or in small groups.

I once dived to a certain deep rock face to experience again the pleasure of seeing a lush sea-anemone garden I had discovered only a week before. I was disappointed; I found a sparse bed of somewhat ragged specimens. After the dive, I went back to my textbooks to learn how these apparently sedentary animals could move into and out of an area so quickly (it seemed unlikely that they could have died off in such numbers in so short a time). I learned that these plantlike animals are capable of locomotion, in some cases by means of rather dramatic and unexpected movements. Sea anemones are reported to be able to change their locations by slow gliding on their foot pads (pedal disks), by crawling on their sides, or by bending over and walking on their tentacles. One West Coast species found in Puget Sound has been observed to move by performing a series of cartwheels. When disturbed, it bows its crown of tentacles down to the rock and then stands on its "hands" as it swings its foot pad through an arc to land in a new spot inches away. It can repeat this unlikely maneuver until it has satisfied its wanderlust or, more importantly, until it has escaped a nearby predatory sea star.

The sex life of sea anemones is not nearly as interesting as their locomotion, but it is worth a few words. These beautiful organisms reproduce sexually, giving rise to free-swimming larvae that eventually settle down to a sedentary existence. Sea anemones can also reproduce asexually, by budding. Occasionally I found a sea anemone with what appears to be a small cut near the margin of its foot pad. When I returned days or weeks later, I saw a miniature sea anemone budding from this spot. Ultimately, a full-grown individual develops and separates from the parent animal, a form of reproduction termed pedal laceration.

Large beds of sea stars are also a striking feature of deep offshore bottoms. These echinoderms typically live on gravel or rock bottoms at all depths to which we dive, but only in deep offshore waters have I seen them grow to truly giant size— individuals ten to twelve inches across are not uncommon. Also, they grow here in great numbers. It is not unusual to see a deep rock bottom paved with the common sea stars *Asterias forbesi* and *Asterias vulgaris* (8, 9, 12). Red, brown, purple, yellow, and white individuals may be so crowded together that their arms everywhere touch or overlap.

At these depths there are also sea stars that cannot be found in shallower water in New England: the bat star, *Patiria miniata* with a pronounced web between each pair of its arms; and the spectacular sun star, with up to fourteen arms. The grotesque basket star *(Gorgonocephalus agassizi)*, a relative of the sea star, is an inhabitant of the deepest water that divers in New England dare to penetrate. Its branched and coiled arms, as its scientific name suggests, resemble a Gorgon's snaky head.

NEW ENGLAND WATERS

CAMERA BELOW

by Jeffrey L. Rotman

TAKING PHOTOGRAPHS beneath cold waters is both fun, and at times, frustrating. Working in this environment with a camera teaches you to search, to see, and above all to be patient. Slowly the ocean reveals its hidden residents; some are elusive because they are small, others because they are well camouflaged, and a few only appear during the winter months when frigid water temperatures had previously cooled my enthusiasm for exploring the ocean. After spending one hour in 30°F water I felt more like an ice cube than an underwater photographer. But the discomfort was forgotten when the film was developed. I was collecting a portfolio of marine behavior that few people had seen. The animals were exquisitely beautiful, and their colors and textures were unique and unexpected.

As I continued to explore the cold water environment along the east and west coasts of North America I found striking parallels although each region supported a different collection of marine life. Jeweled nudibranchs, moon snails with insatiable appetites, and flowerlike anemones, to name just a few, were common to all the cold water environments I visited. Yet each area had its unique forms as well. New England could claim its bizarre goosefish, California its forests of giant kelp, and the Pacific Northwest a sea pen many times larger than any I had seen elsewhere.

Photography in these waters presented problems. When behavior was the subject, patient waiting was required. Movement helps to keep a diver warm, yet it is important to remain almost motionless. I began to rely more and more on a drysuit. Although this suit provided the extra warmth necessary for comfort, its added bulk made moving through the water more difficult.

In addition to the numbing cold, there are low levels of natural illumination. Below 15 to 30 feet (roughly 5 to 10 m.) most of the warm colors—reds, oranges and yellows—are absorbed by the overlying water, leaving the cool greens, blues and violets. To dispel the darkness and restore the lost hues, supplemental lighting must be used. The rich plankton blooms and the fresh-water runoff of the coastal waters of North America make artificial lighting a tricky proposition because certain lighting angles must be used to avoid highlighting particles suspended in the water. The giant kelp forests in California add to these difficulties by screening out even more sunlight, but in doing so create an environment as lush as any jungle.

Technically, just how can these physical and technical obstacles be overcome to capture photographs beneath cold seas? What camera, lens, films, light sources, and accessories should be used? Many underwater photographers use an amphibious camera called the Nikonos. This camera is so versatile that most manufacturers of accessories for underwater photography specifically design their products for it. A deceptively simple viewfinder camera that uses 35 millimeter film, the Nikonos can be submerged without a separate, protective housing and can handle most underwater photographic needs. Another type of camera often used for underwater photography is a single-lens reflex (SLR) camera with the 35 millimeter film format. Nikon and Canon produce popular models. These cameras are designed for abovewater use, so they must be protected with a special pressure-resistant housing when underwater. In addition, they are fitted with a special viewfinder that magnifies the through-the-lens image.

I use both the amphibious Nikonos and a housed 35 millimeter format SLR, because each of these underwater cameras has special advantages. The Nikonos is rugged, compact, and versatile. An SLR will accept a wide range of lenses as well as a motorized film-advancing device. For my SLR, a motorized Nikon, I have an aluminum housing because of the abuse it takes when I am on professional assignment; but plastic housings (especially those of polycarbonate plastics such as Lexan) can be quite good and are considerably less expensive.

Lenses must be selected carefully for underwater work. While the standard lens for abovewater photography in the 35 millimeter film format has a focal length of 50 millimeters, the same angle of view is achieved underwater with a lens of 35 millimeters, because water increases the focal length of abovewater lenses. Long focal length lenses are employed to take frame-filling closeups of tiny subjects. Macrophotography has revealed much of the fascinating small worlds that exist in the cold-water environment. Lenses with a wider angle of view are also used underwater. The wide angle captures a fairly large subject without placing the camera so far away that all detail is obscured by turbid water. An extreme example of an ultra-wide angle lens is the "fish eye," whose circular field and distorted image convey little of reality but produce a striking visual effect.

The amphibious Nikonos may not have a built-in exposure meter, and the special viewfinder often used with a housed SLR may lack the lightmeter incorporated in the conventional viewfinder it replaces. In both of these cases a separate meter must be used: one manufactured specifically for underwater use or a conventional meter protected in a housing. I favor the

CAMERA BELOW

amphibious type, partly because housed systems tend to be bulkier and heavier and housed systems—lightmeters, cameras, or accessories—often contain more complex mechanisms, an added risk in the event of flooding.

Electronic flash is a necessity for underwater photography. Amphibious strobes are favored over their housed counterparts because they are more powerful for their size and weight. Multiple strobes are common, with two-unit setups most frequently employed. The additional strobes are not necessarily mounted on the camera. I prefer to have my diving partner hold a "slave," a self-contained unit with a photocell triggered by light from the "master" strobe at the camera. The slave is held off to one side of the underwater scene to prevent light reflected by suspended particles from reaching the camera. Slaves are generally used for fill-in lighting.

Some photographers use larger film formats underwater, employing cameras such as the Hasselblad and Rolleiflex. These cameras, with their bulky underwater housings, increase the quality of the photographic image because of the larger film area. Excellent results can be obtained with this equipment. However, the advent of powerful electronic flash units for underwater work has made it possible for photographers to use relatively slow films, with their finer grain and higher resolution, so that superb images can be obtained with 35 millimeter equipment. I prefer Kodachrome 64, but Ektachrome 64 is also popular. High-speed films are usually reserved for work in available light or for distant subjects. Cold waters are often dark waters, and a high-speed Ektachrome such as ASA 160, used with available light, can help capture the mood of cold-water diving.

Diving skill is important, too. An amateur photographer with considerable diving experience is likely to do a better job

underwater than a topside professional who dives infrequently. An understanding of subsea environments and an ability to move comfortably and safely through them are prerequisites to the mastery of even the simplest underwater photographic techniques.

After I discovered that the technical and physical challenges of photography beneath cold waters could be met, I had to answer to the animals, the subjects of my work. How was I to convince a small, timid toad crab that I meant no harm? With my two-hundred-pound frame, another one hundred pounds of equipment strapped to it, my smile masked by the mouthpiece in my teeth and an ever-present gurgling of air bubbles made me a formidable intruder. Offerings of food occasionally bought my way into an otherwise closed circle of friends. When I had achieved this, I could move a camera into position. Marine animals sense impending danger and an aggressive approach almost always puts them to flight or at least alters their natural behavior. A slow approach buys the time needed to observe an animal and its habits, and to compose a photograph that captures a special moment. Some of these moments have been preserved in this book.

COLOR PHOTOGRAPHS

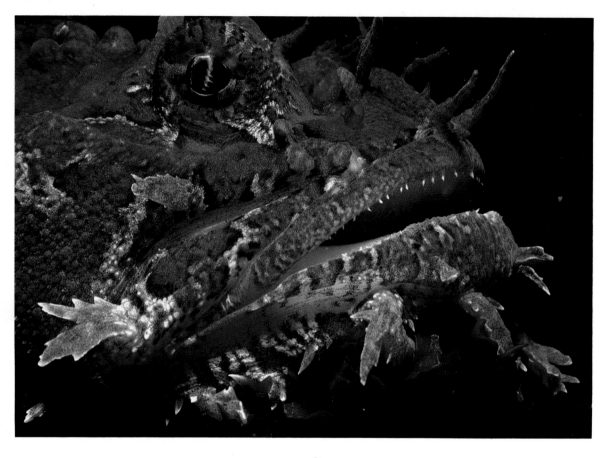

1

Above: Bearded visage belongs to a sea raven, Hemitripterus americanus. *This individual exhibits an uncommonly bright coloration; most are a muted brown. Its fleshy beard blends into the New England coastal environment, especially in the intertidal zone, where beds of the red algae called Irish moss,* Chondrus crispus, *cover the rocks. Although primarily a scavenger, the sea raven sometimes snaps at small passing prey. Gloucester, Massachusetts; 26 ft. (8 m.)*

2

Opposite, above: A sea raven was well camouflaged in a bed of Metridium *sea anemones until my electronic flash brought color to this scene. On a sunken submarine, the U-853, Block Island, Rhode Island; 115 ft. (35 m.)*

3

Opposite, below: Flatness is his disguise. A blackback flounder, Pseudopleuronectes americanus, *sometimes called winter flounder, hides on flat bottoms, exploiting its shape as well as the ability of its pigment-bearing chromatophores to alter its coloration. Cape Ann, Massachusetts; 17 ft. (5 m.)*

4

This blackback flounder is right-eyed. Early in its life it swam upright, with one eye on each side of its head. Then it lay down on its side, and its lower eye, rather than stare forever into the feature-less sand, moved around to join its fellow (in this case the right eye). The adult flounder has the use of both eyes, to detect predator and prey alike. Cape Ann, Massachusetts; 33 ft. (10 m.)

5

Ugly is beautiful. The goosefish, Lophius americanus, *lies camouflaged on the sand to lure unsuspecting fish with the angling appendage carried on its upper lip. Pocketbook mouth and sharp conical teeth make its victims disappear with surprising ease. Cape Ann, Massachusetts; 20 ft. (6 m.)*

6

Put to flight, this immature skate, Raja erinacea, *was surprised at night by my companion's underwater lamp. These fish enter sandy coves in spring and summer to reproduce. Cape Ann, Massachusetts; 63 ft. (19 m.)*

7

Jewellike and no bigger than a thumbnail, a blood star, Henricia sanguinolenta, *is one of the many sea stars that can be found in tide pools by the observant beachcomber. This specimen rests on a background of pink coralline algae and white bryozoans. Cape Ann, Massachusetts; 3 ft. (1 m.)*

8

Caught in the act, a purple sea star, Asterias vulgaris, *preys on a mussel. I arrived just as the sea star embraced its victim and began to apply the slow pull that would breach its defenses. Thirty minutes later, the sea star everted its stomach between the parted shell and began to digest the mussel within its own house. Cape Ann, Massachusetts; 23 ft. (7 m.)*

9

*Born survivors, purple sea stars are able to regenerate lost arms.
This individual is in the process of replacing three of its five
appendages. Pemaquid Point, Maine; 33 ft. (10 m.)*

10

Opposite, above: A brace of periwinkles, Littorina littorea, *file across a common sand dollar,* Echinarachnius parma. *The five-armed star on the sand dollar's circular test tells of its kinship to the sea stars. This flat animal inhabits shallow sandy areas, where it may be so numerous as to pave the bottom. Cape Ann, Massachusetts; 16 ft. (5 m.)*

11

Opposite, below: A pincushion of spines protects this green sea urchin, Strongylocentrotus drobachiensis, *from predators as it grazes on kelp. It clings to the algae with its tube feet, which are visible. Other echinoderms, including sea stars and sea cucumbers, also possess tube feet. Cape Ann, Massachusetts; 26 ft. (8 m.)*

12

Above: Side by side, two purple sea stars lie on a rock encrusted with pink coralline algae. These individuals show only two of the many color variations seen in this species. Cape Ann, Massachusetts; 40 ft. (12 m.)

13

Opposite, above: Offense is the best defense. A rock crab, Cancer irroratus, *strikes a threatening posture as I move in to take its photograph. This common inhabitant of rocky bottoms lives from the intertidal area down to a depth of thirty feet. It scavenges the bottom, eating what others leave behind. Cape Ann, Massachusetts; 13 ft. (4 m.)*

14

Opposite below: A burst of bright bubbles crowns a diver who has captured a good-sized Jonah crab, Cancer borealis, *in the chilly springtime waters of New England, whose temperature is 45°F (7°C). The diver is well dressed against the cold, wearing a thick wetsuit, boots, gloves, and hood. Gloucester, Massachusetts; 33 ft. (10 m.)*

15

Above: Formidable in its armor and with sharp spines bristling from head and claws, the North American lobster can repel many a potential predator. Cape Ann, Massachusetts; 50 ft. (15 m.)

16

Above: A tiny dazzling marvel, this salmon-gilled nudibranch, Coryphella salmonacea, *glides along the just-parted valves of a horse mussel*, Modiolis modiolis, *grazing on minute algae and hydroids growing on the shells. The nudibranch is less than an inch long. Boston Outer Harbor, Massachusetts; 50 ft. (15 m.)*

17

Opposite, above: You are what you eat. The coloration of many marine animals is affected by their diet. For example, nudibranchs feed on a variety of organisms, including sponges, hydroids, sea anemones, bryozoans, and algae, and individuals of this species vary in color depending on their location and the pigmentation of

their primary food source. Halfway Rock, Massachusetts; 50 ft. (15m.)

18

Opposite, below: Rolled up like freshly laundered socks, these three common anemones, Metridium dianthus, *share a rock below the New England thermocline. These anemones are in a contracted state, having expelled much of the water that was in their bodies, but they can reinflate themselves and assume their flowerlike shape. Individuals of this species may be white, orange, chocolate brown, or an intermediate shade. Gloucester, Massachusetts; 83 ft. (25 m.)*

19

Deadman's fingers grope the waters. Red sponges, Haliclona oculata, *with this rather macabre common name share the favorable environment provided by a pair of wooden pilings with a common sea anemone and a cluster of horse mussels. These different animals all feed on plankton brought by the continual movement of the tides. Cape Ann, Massachusetts; 26 ft.(8 m.)*

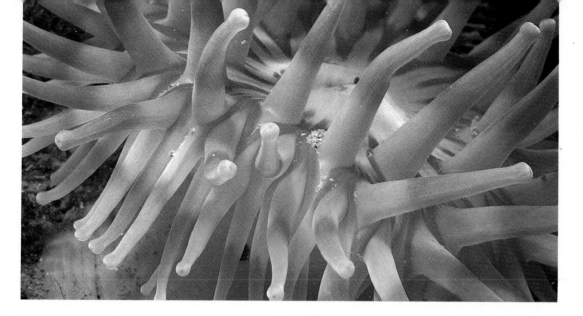

20

Above: *Flower-of-the-sea aptly describes this northern red sea anemone,* Tealia felina. *Cape Ann, Massachusetts; 66 ft. (20 m.)*

21

Below: *Abstract art? Its hollow body turned inwards and its mouth nearly closed, a common sea anemone becomes a study in color, form, and line. Cape Ann, Massachusetts; 83 ft. (25 m.)*

22

With alien grace, this lion's-mane jellyfish, Cyanea capillata, rhythmically contracts its bell, pumping itself through the liquid stillness in search of planktonic food. Its filamentous tentacles are armed with a sufficient quantity of powerful stinging organs, nematocysts, to present a real danger to swimmers. Cape Ann, Massachusetts; 33 ft. (10 m.)

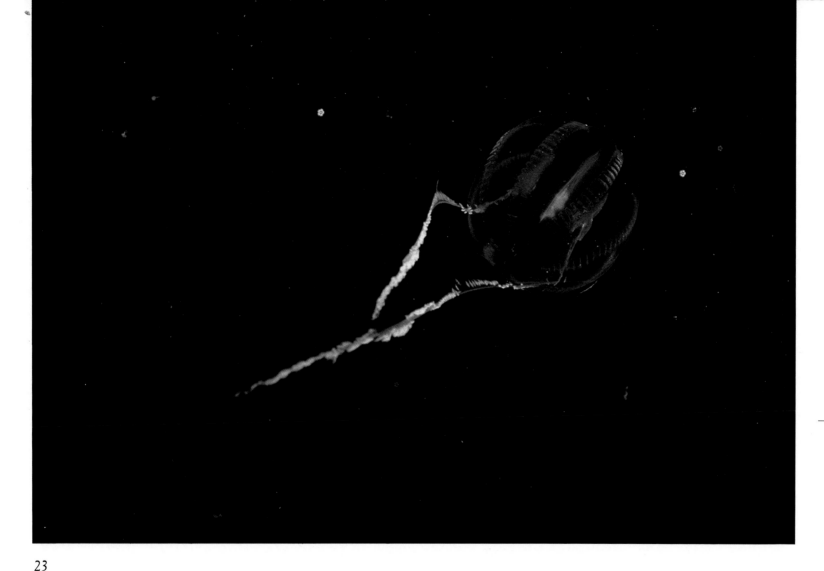

23

Iridescent bands of color are the rows of whiplike flagella, termed comb plates, that propel this translucent comb jelly or ctenophore, Pleurobranchia pileus, through the cold winter waters of New England. The pair of trailing tentacles is covered with a sticky substance to catch tiny plankton, especially fish eggs, upon which this animal feeds. Comb jellies may be so abundant in the winter and early spring that the water swarms with them. The marine biologist, Henry B. Bigelow, wrote about these animals: "Whenever these ctenophores swarm, they sweep the water so clean and they are so voracious that hardly any smaller creatures can coexist with them." Gulf of Maine, Massachusetts; 33 ft. (10 m.)

24

Sea butterfly or pteropod, Clione limacina, *normally lives in deep waters well offshore but at night comes to the surface to feed. It is occasionally seen in coastal waters of the Atlantic in the spring. This individual, only one inch long, was photographed just days after the great blizzard of February 1978, which probably blew it in towards shore. Technically snails, but lacking shells in adult life, these graceful animals are often found in large numbers in pelagic waters, where they are eaten by baleen whales. Gulf of Maine, Massachusetts; 3 ft. (1 m.)*

25

Death for the weak, food for the strong: feeding on its own kind, one northern moon snail, Lunatia heros, *drills its toothed tongue, or radula, throught the shell of another. I arrived only moments before, and watched in astonishment as the smaller snail overwhelmed the larger, engulfing it with its oversized foot. Gulf of Maine, Massachusetts; 26 ft. (8 m.)*

26

Gaping wide, horse mussels use hairlike cilia to propel a current of water over their internal gills, which are modified to filter planktonic food. The shells of the mussels are encrusted with pink coralline algae. Although the purple sea star is this mollusk's natural predator, the immature individual seen here is too small to be a threat. The green sea urchin frequently shares living space with horse mussels. Cape Ann, Massachusetts; 73 ft. (22 m.)

27

Armored *against its enemies with eight overlapping plates and with
its speckled green mantle spread like a full skirt over the rock, a
chiton,* Ischnochiton ruber, *grazes peacefully on algae. This ani-* *mal clings so firmly to the rock that it is almost impossible to
dislodge. Cape Ann, Massachusetts; 30 ft. (9 m.)*

28

A *living fossil: the horseshoe crab,* Limulus polyphemus, *has survived nearly unchanged through 600 million years of evolution and, like the individual seen here, is commonly found on sandy bottoms in the spring and early summer. These animals are misnamed; they are closer to spiders than to crabs. Cape Ann, Massachusetts; 23 ft. (7 m.)*

29

Ménage à trois. A train of horseshoe crabs chugs through the sand. During the spring mating season it is not uncommon to see a female with one or two males in tow. At the highest monthly tide, the gravid female deposits her eggs in the sand. The male (or males) ejaculates sperm over the nest. About a month later the tide rises again to this point, releasing the new hatchlings. Gulf of Maine, Massachusetts; 23 ft. (7 m.)

30

Sea peach is the common name for the tunicate Halocynthia pyriformis. *This one is a bright splash of color on the New England hard bottom. These sedentary filter feeders become quite common below thirty feet (ten meters.) Cape Ann, Massachusetts; 75 ft. (25 m.)*

31

The moment captured here was not fully appreciated until this photograph was processed, enlarged and contemplated. If you look closely you will see two planktonic amphipods curled about the body of a third, possibly for the purpose of reproduction. These animals are less than half an inch long (10 mm.). Cape Ann, Massachusetts; 13 ft. (4 m.)

32

Dining on a dead claw, these two moon snails drill their toothed radulas through its outer shell to reach the meat inside. This lobster claw may have been jettisoned in battle, a defensive maneuver common to a number of marine invertebrates. Although moon snails do feed on live prey, they are usually scavengers. Cape Ann, Massachusetts: 23 ft. (7 m.)

33
Cousin against cousin: a blood star feeds on a sea urchin. Both
creatures are echinoderms, or spiny-skinned animals. Cape Ann,
Massachusetts; 20 ft. (6 m.)

34

Discretion is the better part of valor. An unequal match pits a large hermit crab, Pagurus pollicaris, *inhabiting the abandoned house of a dead moon snail, against a smaller member of the same species,* wearing a periwinkle shell. The smaller crab eventually took the only reasonable course of action—retreat. Cape Ann, Massachusetts; 43 ft. (13 m.)

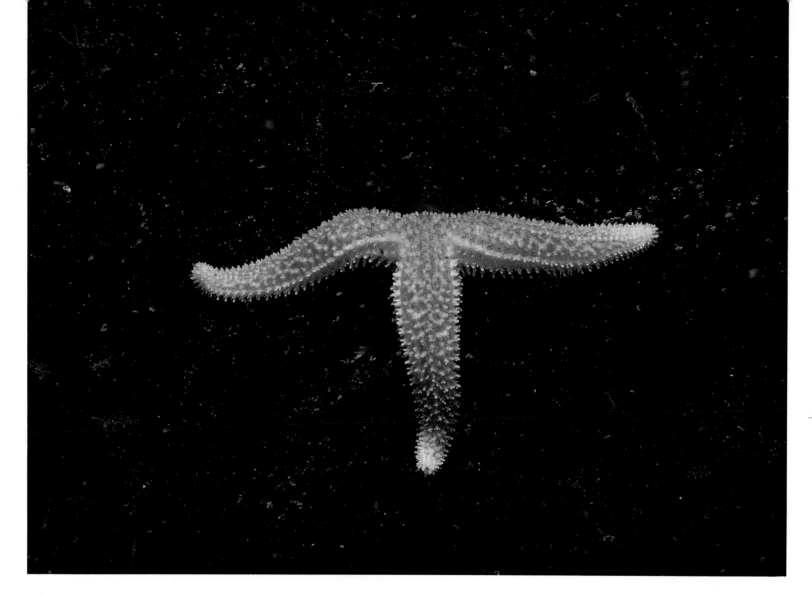

35
The natural symmetry of this sea star is starkly accentuated as it transits a blade of kelp. The spare simplicity of this moment seems to capture the essence of all plant and animal life. Cape Ann, Massachusetts; 33 ft. (10 m.)

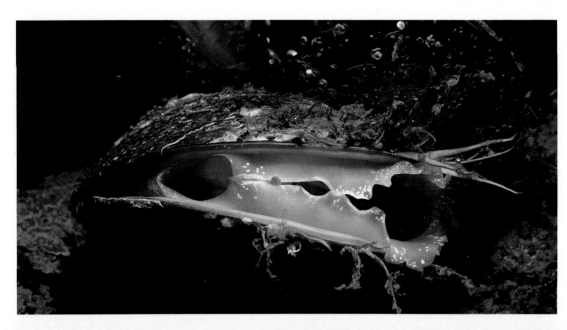

36

Opposite, above: Puckered to channel outward currents, the white mantle of a horse mussel visibly aids this animal's filter feeding. The shells are firmly anchored to the rock by the byssus, a tuft of stout threads. Cape Ann, Massachusetts; 50 ft. (15 m.)

37

Opposite, below: Headless and defenseless, this crab has fallen prey to a quartet of hungry cunners, Tautogolabrus adspersus. These fish frequently scavenge the bottom, helping to keep the ocean in ecological balance. Cape Ann, Massachusetts; 26 ft. (8 m.)

38

Above: Barnacles hold no interest for this blood star; the fibrous yellow sponge provides the nourishment needed for life to go on. Cape Ann, Massachusetts; 20 ft. (6 m.)

39

Face to face, this North Atlantic lobster clamps the rim of its captor's face mask in a futile gesture of defiance. Cape Ann, Massachusetts; 33 ft. (10 m.)

40, 41

Hitchhiker aboard, a female horseshoe crab tows one or more suitors across the bottom for weeks. The reason for this seemingly tolerant behavior is to insure external fertilization of her eggs as soon as they are deposited. Gulf of Maine, Massachusetts; 83 ft. (25 m.)

63

42

Opposite, above: Unexpected riches delight human visitors beneath cold waters. Encapsulated within a drysuit, the diver is well protected from the 48°F (9°C) water as he explores the wall of an underwater seamount carpeted with a lush growth of corals and sea anemones. Southern California; 100 ft. (30 m.)

43

Opposite, below: A plume of silvery bubbles is exhaled by a diver squeezing between giant boulders to examine yellow sulfur sponges, Verongia thiona, *pink strawberry club anemones,* Corynactus californica, *and giant-spined sea stars,* Pisaster giganteus. *Southern California; 90 ft. (27 m.)*

44

Above: Night feeders, strawberry club anemones hungrily sweep the water with their tentacles. At night, Santa Catalina Island, California; 26 ft. (8 m.)

45

Opposite, above: Suffused with green light, a grove of giant kelp, Macrocystis pyrifera, *shelters a variety of animals, including the sunflower star,* Pycnopodia helianthoides, *possibly the world's largest species of sea star. Sunflower stars may have as many as twenty-four arms. Santa Rosa Island, California; 50 ft. (15 m.)*

46

Opposite, below: Wrenched from their rock in a storm, the interlaced holdfasts of a bundle of giant kelp are adrift in midwater, *supported by the buoyancy of the plants' bladders. A sunflower star clings to the uprooted algae. Santa Cruz Island, California; 59 ft. (18 m.)*

47

Above: A constellation of sea stars feeds on a bed of mussels. Empty shells testify to the voracity of these predatory echinoderms. Southern California; 100 ft. (30 m.)

48
Upturned rays of two sea bats, Patiria miniata, *are part of an unexplained behavior that frequently occurs when two of these animals meet. Southern California; 33 ft. (10 m.)*

49

Warm-blooded in cold waters, the California sea lion, Zalophus californianus, is perhaps the most playful, friendly animal in the sea. On numerous occasions I have been treated to a display of speed and grace as these air-breathing mammals would dive, circle and pirouette before my lens. Santa Catalina Island, California; 10 ft. (3 m.)

50, 51, 52, ,53

Writhing in flight, this beautiful nudibranch, the Spanish shawl,
Flabellina iodinea, *swims by doubling and twisting its body, as
shown by this series of photographs taken at twenty-second inter-
vals. The bright pigmentation of this and most other nudibranchs is
thought to signal potential predators of the poisonous nature of
these small mollusks. Southern California; 36 ft. (11 m.)*

54

Opposite: Aggressive hustlers, these flashy garibaldis, Hypsypops rubicunda, *came running when my companion opened a sea urchin with his knife. The garibaldi has enjoyed protection from collectors and spear fishermen since it became California's state marine fish. Santa Catalina Island, California; 40 ft. (12 m.)*

55

Above, left: Doubled rings of tentacles are characteristic of this tube anemone, Cerianthus aesturi; *it extends its tentacles from a parchmentlike tube anchored in the sand as it sweeps the water for plankton. Southern California; 16 ft. (5 m.)*

56

Above, right: A field of marigolds is brought to mind by this bed of orange cup corals, Balanophyllia elegans. *These sedentary animals feed by fanning the water with their tentacles, capturing tiny planktonic prey. The corals have grown in and around an encrusting cobalt sponge,* Hymenamphiastra cyanocrypta. *Cup corals prefer shady spots and often cluster beneath overhangs. Santa Catalina Island, California; 26 ft. (8 m.)*

57

Opposite: Large and lumbering, a spider crab, Pugetta producta, *was ascending a bundle of giant kelp when its journey was interrupted. It was captured by hand and examined eyeball to eyeball. Southern California; 50 ft. (15 m.)*

58

Above: Vanity is not his motive; this decorator crab, possibly Loxorhynchus crispatus, *uses rows of tiny hooks, called setae, on its carapace and appendages to attach bits of sponge and other suitable materials to its exoskeleton. The borrowed garb may be more than camouflage: in aquariums, crab-eating fish have been observed to take individuals decorated with stinging hydroids and promptly spit them out. These crabs molt under cover of darkness and quickly decorate their new shell. Members of this species abandon the decorating habit at maturity, when they are presumably large enough that they no longer must hide from their enemies. Southern California; 59 ft. (18 m.)*

59

60

Opposite: Sunbeams, kelp, and the blue empyrean provide the backdrop for a surfacing diver clutching a spiny lobster, Panulirus interruptus. *Southern California; 26 ft. (8 m.)*

Above: Bristling with spikes, the head of this spiny lobster is equipped with sharp projections that are a formidable defense. At night, Southern California; 33 ft. (10 m.)

61
Hard to hold, the spiny lobster is no easy catch. It can strike a smart backhand blow with its powerful tail, driving sharp spines into a predator—human or otherwise. Southern California; 33 ft. (10 m.)

62

As if in a field of strawberries, a barnacle, Balanus tintinnabulum, *fans in a planktonic meal with feathery tendrils, surrounded by strawberry club anemones. Southern California; 23 ft. (7 m.)*

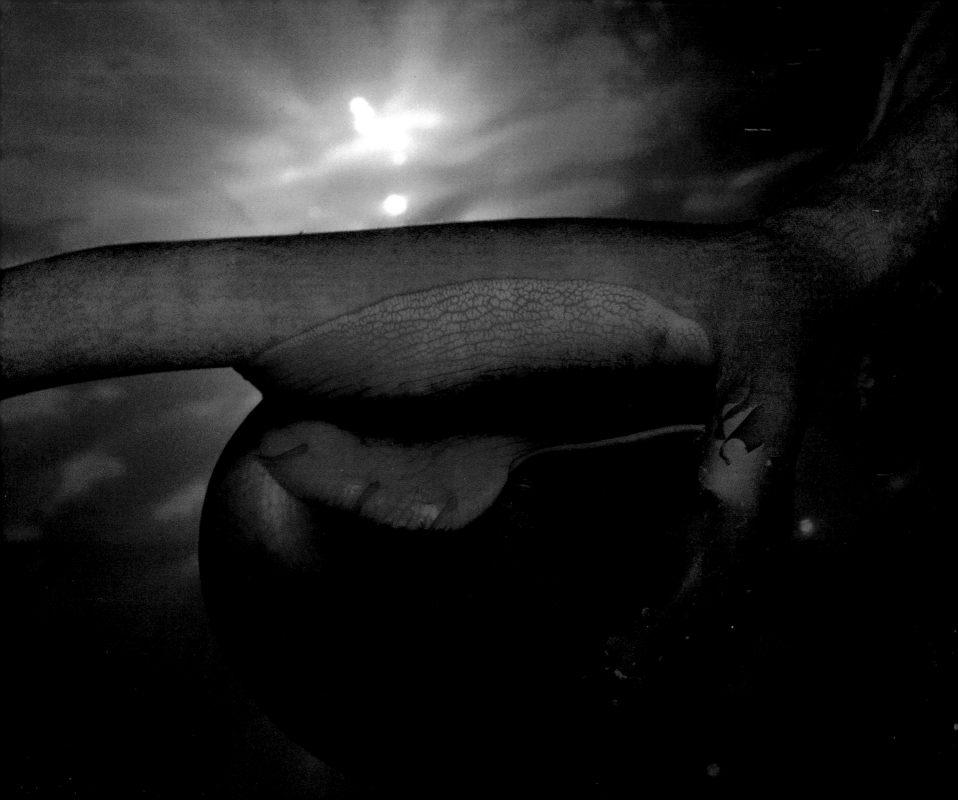

63

Opposite: this two-inch-long kelp snail, Norrisia norrisi, *grazes its way up the full length of a kelp plant, removing a thin layer of cells along the way. When it reaches the top, it releases itself to free-fall to the bottom, where it begins its food-gathering journey once again. Santa Catalina Island, California; 33 ft. (10 m.)*

64

Above: In strange communion, diver and fish are silhouetted among the kelp. Santa Rosa Island, California; 50 ft. (15 m.)

65
*Sleeping angel: believing its camouflage effective, a Pacific angel
shark,* Squatina californica, *lies motionless in the sand. Santa
Catalina Island, California; 40 ft. (12 m.)*

66

*Coming out of the woodwork, this fringehead fish, Neoclinus
uninotatus, is marvelously camouflaged as it peers from its hole. Its
head is the size of a pencil eraser, and the fleshy fringe for which it is
named helps it blend with the algae around it. My discovery of this
retiring animal was a stroke of good luck. Monterey, California;
26 ft. (8 m.)*

67
It goes where it pleases: the Pacific electric ray, Torpedo califor-
nica, *is capable of producing an electric shock of up to eighty volts.
Santa Rosa Island, California; 33 ft. (10 m.)*

68

Gossamer pectoral fins fan the water as this colorful rockfish, a species of the genus Sebastes, hovers warily above a carpet of algae, sponges, and sea anemones. Relying on its camouflage, the fish lay motionless as I moved my camera lens to within four inches of its head. Rockfish abound on West Coast reefs and include at least sixty species, many of which are difficult to distinguish since they differ only slightly in coloration. Santa Cruz Island, California; 33 ft. (10 m.)

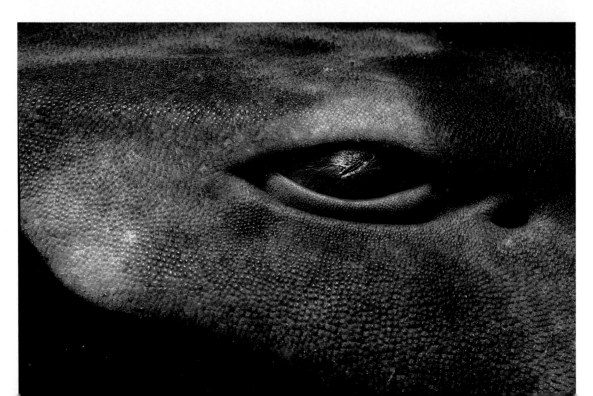

69

Opposite, above: Fearsome rows of inward-pointing teeth line of the mouth of a California swell shark, Cephaloscyllium ventriosum, but this retiring bottom dweller is generally no hazard to humans. These sharks usually inhabit crevices or caves in shallow bottoms; when disturbed, they inflate themselves with water, the behavior for which they are named. Southern California; 50 ft. (15 m.)

70

Opposite, below: Slitlike pupil belongs to the eye of a California swell shark. To capture this photograph, we had to restrain the shark, not a very comfortable experience for it or for my assistant.

The fish continuously tried to turn and bite the hands that held it. Just before a shark attacks it moves a protective membrane over its eye; this behavior is illustrated here. Santa Catalina Island, California; 40 ft. (12 m.)

71

Above: I searched in vain for this animal in daylight, but when I returned at night, I saw six of these lovely chestnut cowries, Zonaria spadicea, in less than an hour. This inch-long specimen grazes on a red volcano sponge. At night, Santa Catalina Island, California; 33 ft. (10 m.)

72
Human hunter patrols the perimeter of a kelp grove, hoping to find a halibut or sea bass. Either of these fish can exceed 100 pounds (45 kilograms). Southern California; 66 ft. (20 m.)

73

Bright branching fingers of coral such as these California hydrocorals, Allopora californica, are smaller and less common than the corals of the tropics, but no less beautiful. When the individual polyps are extended, their minute tentacles capture and paralyze nearly invisible planktonic animals. It may take twenty years for a colony to reach a height of twelve inches (thirty cm.). Though corals are found in these and other cold-water regions, true reef-building corals occur only in warm wters. Southern California; 59 ft. (18 m.)

74

Opposite: Furrowed by a light swell, the Southern California kelp mat catches the glancing rays of the setting sun. Santa Catalina Island, California; surface.

75

Above: Feeling its way along the night bottom, this sea bat is a nocturnal predator, as are many other sea stars. Northern California; 33 ft. (10 m.)

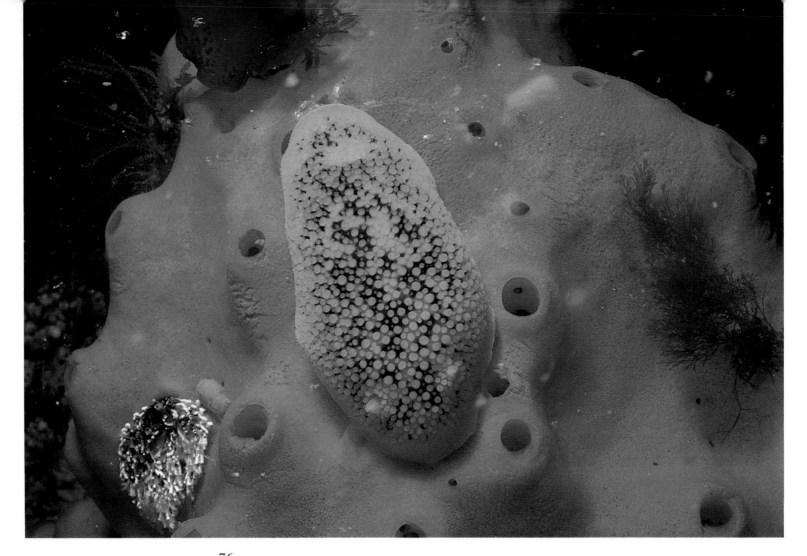

76

Fire and brimstone are the hot colors of this scene. A lemon nudibranch, Anisodoris nobilis, *feeds on a volcano sponge,* Acarnus erithacus, *and partly owes its coloration to its prey. This species of nudibranch ranges from Vancouver Island to Laguna Beach. Monterey, California; 40 ft. (12 m.)*

77
Ensuring its posterity, a ringed nudibranch, Discodoris sandiegensis, *extrudes its spiral egg ribbon. This subtly colored animal ranges from Unalaska Island to San Diego. Monterey, California; 33 ft. (10 m.)*

78

Above: Living in borrowed quarters, the Bering hermit crab,
Pagurus beringanus, *is frequently seen in the intertidal zone, where it scavenges a meal or searches for a larger shell. Puget Sound, Washington; 10 ft. (3 m.)*

79

Opposite, above: Fanning for plankton, this giant barnacle, Balanus nubilus, *measures less than two inches (four centimeters) in diameter at its base. A young sun anemone,* Metridium senile,

competes with the barnacle for the same planktonic meal. Puget Sound, Washington; 33 ft. (10 m.)

80

Opposite, below: Bright sunburst in an emerald firmament, this sunflower star belongs to a species that is known for its voracious appetite, sometimes even devouring its own kind. It begins life with six rays and at maturity can have as many as twenty-four. Vancouver Island, British Columbia; 43 ft. (13 m.)

81

Tightly contracted, a white-spotted rose anemone, Tealia lofoten-sis, *has expelled much of the water that keeps it erect. In this state it uses little metabolic energy and rests in relative safety. Monterey, California; 66 ft. (20 m.)*

82

Surreal flower—a diver admires a spectacular sea anemone, a species of Tealia, *that has managed to reach a respectable size. Monterey, California; 83 ft. (25 m.)*

83

Patience rewarded, I watched this sea cucumber, Eupentacta quinquesemita, *as it thrust each of its ten branching arms one by one into its central mouth to be licked clean of captured plankton and organic matter. A full cycle took nearly ten minutes, then began again. This species ranges from Northern California through British Columbia. Northern California; 66 ft. (20 m.)*

84

Disturbed by my presence, the tentacles of this sea anemone are about to disappear. As soon as my first flash went off, the animal began to withdraw its tentacles, a defensive behavior that I captured in this photograph. At night, Monterey, California; 83 ft. (25 m.)

85

"Fear not, till Birnam Wood do come to Dunsinane." Like Macbeth's enemies, this decorator crab conceals its presence with "leafy screens," pieces of its natural surroundings—in this case, sponge and coralline algae. Point Lobos, California; 50 ft. (15 m.)

86
In the somber light of a kelp grove, a bright orange sea star searches the bottom for food. Point Lobos, California; 40 ft. (12 m.)

87

Don't tread on me. The sharp spines of this red sea urchin, Strongylocentrotus franciscanus, *ward off would-be predators. This animal feeds primarily on algae, but is also a scavenger and* will eat most dead organic matter that comes its way. Monterey, California; 26 ft. (8 m.)

88

Sharp spines keep predators at bay as these purple sea urchins, Strongylocentrotus purpuratus, *creep imperceptibly along the bottom, feeding on algae and other organic matter. Santa Rosa Island, California; 23 ft. (7 m.)*

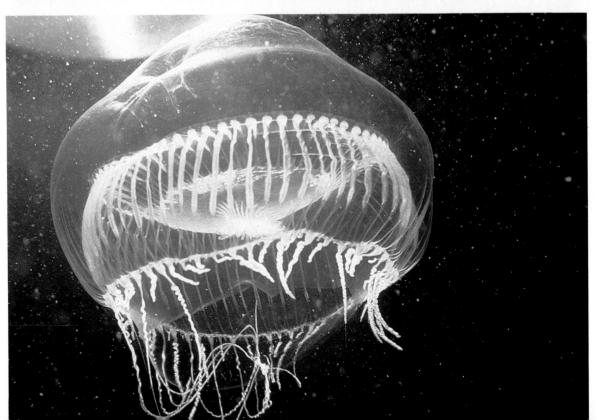

89

Opposite, above: Bloated with water, a red sea cucumber, Parastichopus californicus, *is a common member of the bottom community in the Pacific Northwest. This individual has retracted its feeding appendages into its circular mouth and inflated itself with water, its typical defensive behavior. Puget Sound, Washington; 26 ft. (8 m.)*

90

Opposite, below: Little more than organized water, this jellyfish, Aequorea aequorea, *is only two inches in diameter. It propels itself through the water with rhythmic pulsations of its transparent bell. Puget Sound, Washington; 7 ft. (2 m.)*

91

Above: A Brobdingnagian beauty, this sea pen, Ptilosarcus gurneyi, *was many times the size of any I had seen before. Anchored in the sand bottom and blowing in the swift current, this colonial animal is related to corals and sea anemones. Perhaps the strong current carried a rich supply of plankton and was thus responsible for this individual's immense size. Vancouver Island, British Columbia; 50 ft. (15 m.)*

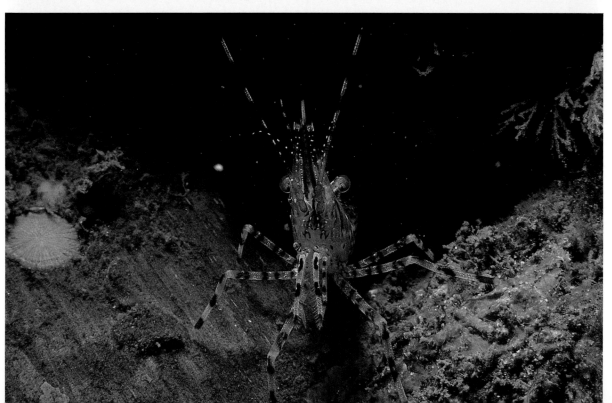

92

Opposite, above: Looking like rock, this Puget Sound king crab, Lopholithodes mandtii, *is as weird as any I have seen in the ocean. Although this specimen could not have weighed more than four ounces, there are authenticated reports of individuals of this species weighing as much as fifteen pounds (seven kilograms). Hornby Island, British Columbia; 40 ft. (12 m.)*

93

Opposite, below: All legs and eyes, a coonstripe shrimp, Pandalus danae, *became irrepressibly curious when I began to take its photograph. As I moved my lens closer, the shrimp advanced to meet it and eventually left its crevice home to climb onto my camera. Puget Sound, Washington; 33 ft. (10 m.)*

94

Above: Like peppermint sticks, the brightly banded tentacles of the Christmas anemone, Tealia crassicornis, *are arranged in concentric circles. This individual has drawn the first circle of tentacles to its mouth in what may be feeding behavior. The same species of sea anemone is found off the coast of California. Vancouver Island, British Columbia; 50 ft. (15 m.)*

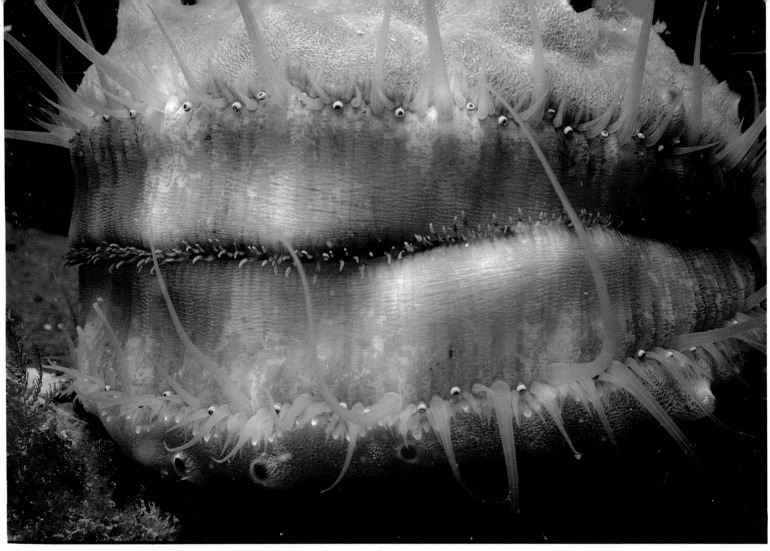

95

Above: *Hundreds of eyes peer out from the white mantle fringes of this scallop,* Chalamys rubida. *If the eyes detect danger, this mollusk claps its valves together and is jetted backwards. The scallop's shells are encrusted with the sponge* Ectyodoryx parasitica. *I have yet to encounter a scallop in the Pacific Northwest that does not carry this symbiotic organism. Puget Sound, Washington; 66 ft. (20 m.)*

96

Opposite, above: *Fallen archangel is one common name for this*

alabaster nudibranch, Dirona albolineata. *This inch-long specimen moves its translucent body over a rock encrusted with coralline algae. Its leafy cerata are limned in white. Puget Sound, Washington; 50 ft. (15 m.)*

97

Opposite, below: *As if to touch, the rays of these sea bats show rich coloration. Sea stars of this species range the entire west coast of North America, but are most common off southern and central California. (Monterey, California; 66 ft. (20 m.)*

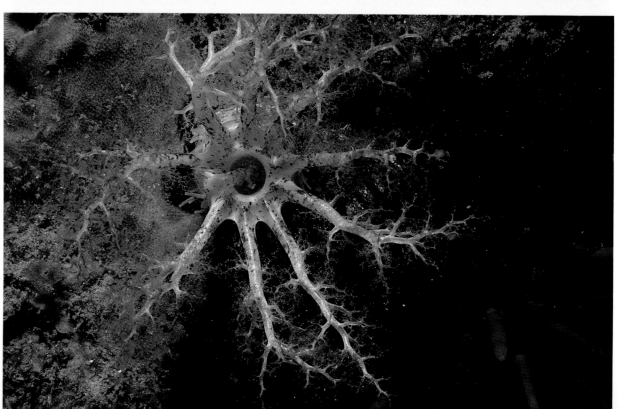

98

Opposite, above: Practically closed, this unidentified species of sea anemone has withdrawn its tentacles and expelled much of the water that gives it shape and structure. Puget Sound, Washington; 50 ft. (15 m.)

99

Opposite, below: Licking its fingers, this sea cucumber sucks an

appendage clean of captured plankton. Hornby Island, British Columbia; 43 ft. (13 m.)

100

Above: Like a coiled spring, the spiral egg ribbon of a mollusk—probably a nudibranch—adheres to a rock encrusted with pink coralline algae. Northern California; 40 ft. (12 m.)

101

Above: Strikingly colored as a warning to predators, this horned nudibranch, Hermissenda crassicornis, grazes on orange cup coral. The coral possesses poisonous nematocysts that the nudibranch consumes and then incorporates into the tips of the horn-like cerata on its own back. This species of nudibranch may be found from Alaska to San Diego. Monterey, California; 33 ft. (10 m.)

102

Opposite, above: Pink calcified algae, a species of Bossiella, is sometimes called wing-nut algae for the appearance of its jointed segments. Here, the algae are festooned with yellow hydroids, Garveia annulata, sedentary colonial animals related to sea anemones and free-swimming jellyfish. Each hydroid possesses tentacles armed with stinging nematocysts capable of paralyzing small planktonic animals. Monterey, California; 23 ft. (7 m.)

103

Opposite, below: Beauty without mercy, this hungry maw belongs to a Christmas anemone, Tealia crassicornis. This cold-water genus has circumnavigated the pole in the Northern Hemisphere and lives on both coasts of North America, extending as far as southern California in the Pacific and Maine in the Atlantic. Monterey, California; 20 ft. (6 m.)

104

Overleaf: Waiting for dark, a diver prepares for a night dive in the chilly waters of the Strait of Georgia. Diving at night holds many surprises and an added measure of adventure. Hornby Island, British Columbia; surface. (Photo by Galit Rotman.)

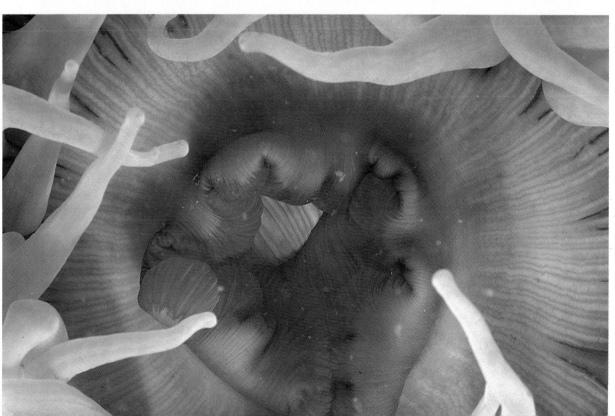

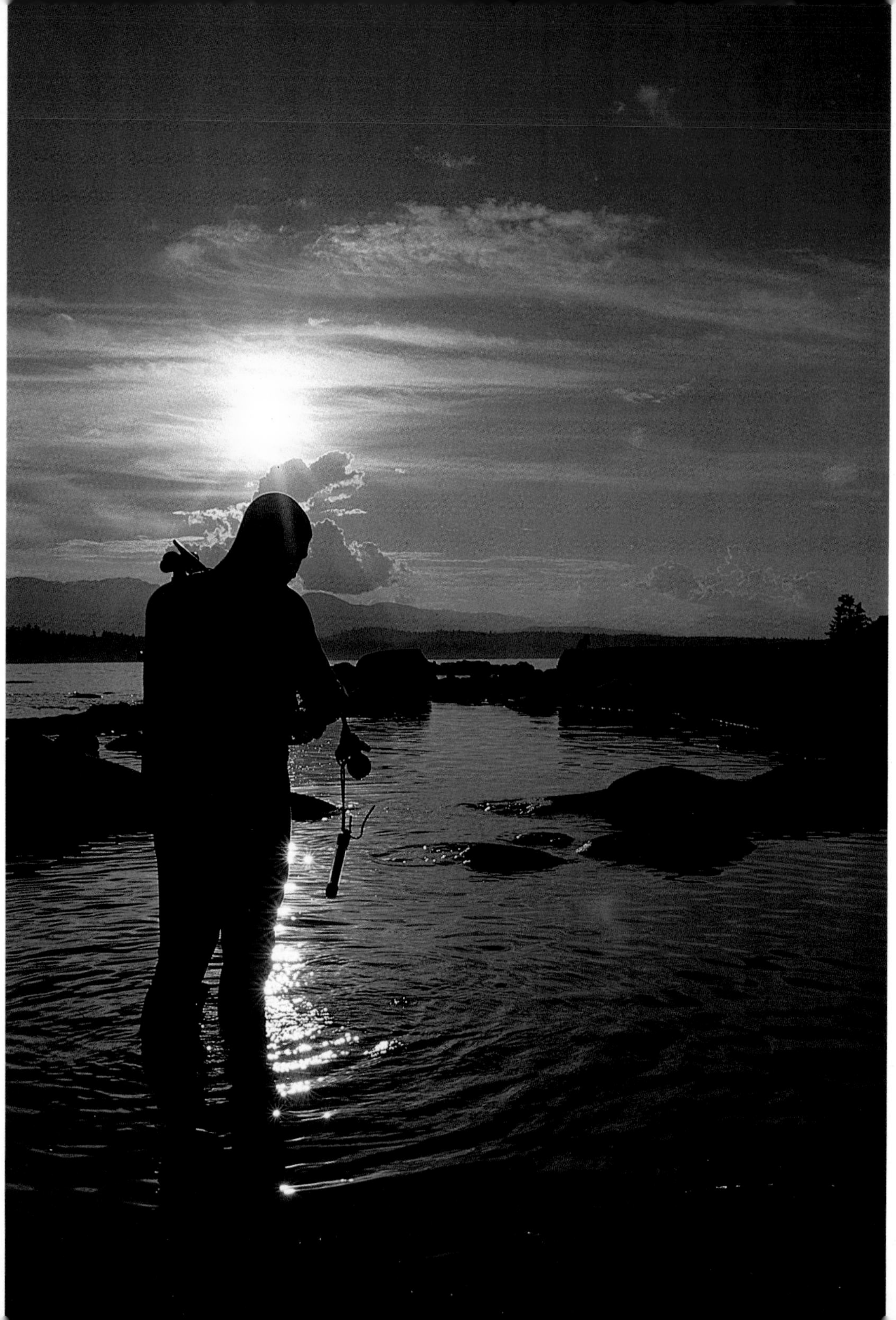

3/ SOUTHERN CALIFORNIA WATERS

California Coasts

CALIFORNIA'S COASTLINE is long and varied. It has broad sand beaches in the south, rock cliffs in the Big Sur, and rolling hills protected by gravel beaches in the north. Surf is characteristic of many West Coast beaches; a steady north-westerly swell moving across the long fetch of the Pacific rolls onto much of the coast. (In New England, big surf usually comes only with storm weather.)

An important geologic difference between the western and northeastern coasts of North America is that the continental shelves are generally narrower in the West. The shelves, actually part of the continents themselves, are the submerged margins of

the land and define the shallow coastal waters. The shelves slope gradually down to about six hundred feet (approximately 200 meters), where there is often a sharp break; here the shelves give way to the continental slope, which drops steeply to the deep ocean floor. The floor is a flat plain of basalt overlaid with the sediment of eons, except where it is deeply cleft by midoceanic rifts or piled up with volcanic mountain ranges (as, for example, the Hawaiian archipelago).

The relatively narrow continental shelves of the West Coast, the prevailing winds, and the earth's rotation all work together to have an interesting and important consequence: steady northwesterly winds blow across the waters overlying these shelves, and because of the earth's spin, water set into motion by wind tends to the right of the wind in the Northern Hemisphere (the reverse is true in the southern latitudes.) Thus, the northwesterlies result in a westward movement of warm surface water out to sea, and colder deep water wells up from below. Since the deeper water comes from beneath the sunlit zones, its rich store of plant nutrients has not been depleted by plant growth. Though upwelling occurs in many parts of the world ocean, the California coast is well suited to produce it intensely and relatively close to shore, since its continental shelf is steep and narrow.

The California current displaces relatively cold water to the south down the West Coast of the United States. Together with upwelling, this cold-temperate current produces a cold-water regime along much of the California coast. For this reason, reef-building corals are not found off Southern California or off Baja California, Mexico, although they are found at nearly the same latitude in the Atlantic off the coast of Florida. In fact, upwelling and cold currents exclude reef-building corals from the west coasts of all the Americas. But there is a unique group of organisms that are characteristic of these waters and that, like the reef corals of the tropics, shelter a rich community of other organisms and thus create an environment that every diver should have the good fortune to experience. These organisms are enormous plants, the giant kelp; the special environment they create is the kelp forest.

Giant Kelp

The fastest-growing multicellular organisms are plants, giant kelp, especially members of the Southern California species

Macrocystis pyrifera (50). Individual members of this species can grow to more than one hundred feet in a single year. Valuable chemicals, including algin and iodine, are extracted from these plants, and an industry has grown up to harvest them. Ships equipped with huge mower blades sheer off the kelp about three feet beneath the surface. The kelp grows back quickly—the harvester comes back for more in a matter of weeks.

Like all algae, kelp are not as complex as land plants. They have no roots, flowers, vascular system, or true leaves. Further, they do not reproduce by means of seeds, but release spores into the water. Seed-bearing plants have specialized tissues for different functions. One kind of tissue comprises leaves, which gather sunlight for photosynthesis; another makes up the stem or trunk, which provides mechanical support; a third becomes the roots that anchor the plant and draw nutrients and water from the soil. Such specialized tissues are said to be differentiated. Algal tissues, on the other hand, are not highly differentiated; the same or similar kinds of tissue serve some of the functions performed in seed-bearing plants by leaves, stems, and roots. The entire plant is considered a single structure, the thallus. The parts of the thallus that have a support function (like the stems of seed-bearing plants) are called stipes, but they also gather light and carry out photosynthesis (like leaves) and absorb water-soluble nutrients from their surroundings (like roots). The flattened parts of the thallus that have the appearance of leaves are the blades, but unlike the true leaves of seed-bearing plants they perform both photosynthesis and nutrient absorption. The algae are anchored to the bottom by means of their holdfasts.

California Islands

Off the California coast between Los Angeles and San Diego are the islands of Santa Catalina, San Clemente, Santa Barbara, and San Nicholas. San Nicholas is the farthest from the mainland. It is inhabited by a colony of northern elephant seals *(Mirounga angustirostris)*. These seals have also been seen on Point Bennett on San Miguel Island, and they are truly magnificent creatures. The males may attain lengths of more than twenty feet (six meters) and weigh up to four tons; females may be half that length and weigh up to one ton. Adult males have an inflatable proboscis for which the species is named. San Nicholas was also once home to thousands of sea otters, but their population was greatly reduced in the 1830s, when great numbers were slaughtered for their pelts. To the northwest of San Nicholas lies Begg Rock, a speck on the map that is known as a magnificent place to dive; its flanks are covered with white sea anemones and enormous rock scallops. Visibility underwater is one hundred feet or greater on occasion; but this is a difficult dive and is hard to reach, as it is far from land and exposed to sea and weather. Of these southerly islands, San Clemente and Santa Catalina are the largest and closest to the mainland, and hence most accessible to divers.

To the north, between Los Angeles and Santa Barbara, is another chain of islands, the northern channel islands of San Miguel, Santa Rosa, Santa Cruz, and tiny Anacapa. These islands fall near the northern boundary of the warm-water province of the temperate zone (the cold-water province begins north of Point Conception), and thus share many underwater plant and animal species with the cold-temperate province.

The waters surrounding both the southern and northern islands are shallow enough—100 feet (30 meters) or less—to be reached by divers, yet because of their separation from the mainland they provide a unique opportunity for the visitor, especially one who dives, to experience a natural beauty that is among the greatest that California, or any other place, has to offer.

San Clemente is a military reservation; besides the military, the island is inhabited by thousands of wild goats. A few dive boats that work out of San Pedro (just south of Los Angeles) take divers to San Clemente; the seventy-eight-foot *Charisma* is one. This island's great sandstone cliffs are deeply pockmarked from naval gunnery practice, and the island has a parched and sterile look. Underwater, however, is a different matter.

We dived off the west-facing shore of San Clemente's southern tip. On this dive I saw two moray eels and large purple sea urchins, *Strongylocentrotus purpuratus*, that were withdrawn into holes, their spines presenting a phalanx to the outside world and presumably to their enemies (88). Other divers speared kelp bass, *Paralabrax clathratus*. I saw a number of neon gobies, small, brilliantly colored bottom fish. In places the gobies were spread roughly equidistantly from each other, as if there were opposing social forces of gregariousness and reclusiveness that caused them to people the bottom in a nearly regular array.

I saw an abalone, a shallow dome of a shell with a curving line of diminishing holes penetrating it near one margin. The holes provide an outlet for the stream of water that the animal maintains for both respiration and feeding. The interior of the shell is beautifully iridescent. The flesh of the abalone's muscular "foot" is prized as food, and the taking of abalone in California is strictly regulated by state law. I thoughtlessly tried to

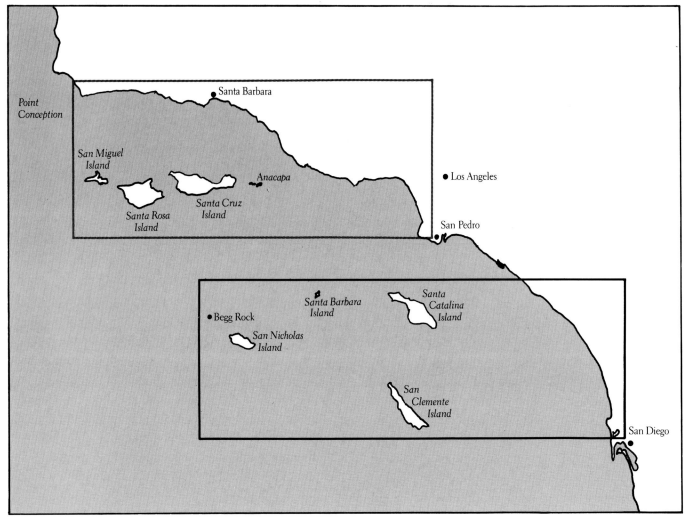

Figures 3.1, 3.2

California's Southern Channel Islands (enclosed by the larger rectangle) lie southwest of Los Angeles. The Northern Channel Islands (enclosed by the smaller rectangle) lie just to the south of the city of Santa Barbara and Point Conception.

pry this abalone off its rock with my diver's knife, a mistake and certainly a violation of California's game laws. I cut its soft foot and a lavender stain grew as its blood diffused through the water.

I turned away and swam on. Many tiny, blue tube anemones covered a rock surface; red gorgonians, a species of coral, looked like garlands of Christmas tinsel. Blackeyed gobies, *Coryphopterus nicholsii*, small sand-colored fish with large dark eyes almost popping out of their heads, clung to the bottom. Short-branched kelp were attached to a rock that loomed above me, and they flowed with the surge; the light faded or grew as the kelp rhythmically swept over me or away. Off in the middle distance a stand of giant kelp evoked a stillness, a feeling of mystery, and provided an enclosing but comforting darkness. A pair of keyhole limpets, *Diodora aspera*, moved slowly over a boulder among potatolike algae, *Colpomenia sinuosa*, that stood on short stalks. Purple lichenlike algae thinly encrusted the rocks, probably a species of the calcareous *Lithothamnium*. I saw a snail on a kelp frond; the orange margin of its foot helped me to identify it as a smooth turban or kelp snail, *Norrisia norrisi* (63).

Santa Catalina lies to the north of San Clemente and is an inhabited island with a city, Avalon (incorporated in 1941). Catalina's year-round population is 1,800, but in the tourist season it may exceed 10,000. There are scheduled boats to Avalon; less ordinary way of getting to the island is by the amphibious Grumman *Mallard* built in 1947 but recently rebuilt and painted a resplendent red, white, and blue. When I flew on the Mallard, I sat in the most forward row of seats and shared the small cabin with five or six other passengers, most of whom appeared to be tourists. The door to the flight deck had been removed so that during the twenty-minute flight I could watch the pilot and copilot and see Catalina move downwards on a tiny radar screen, a light shape against the dark green. I got out of my seat to stand in the aisle and look out the pilot's windshield as Catalina's mountainous skyline rose up through a slight haze. We dropped down through wisps of cloud. The sea was calm but not flat, and the sun sparkled on it. The Mallard and the water met with a deep-felt thud. The pilot played with the throttles as we motorboated across the surface. The brightly painted pontoon that hung down from the port wing just opposite my window alternately split or left the water as the Mallard found its balance. We came to rest at a buoy where a small pontoon boat motored out to meet us. A short trip, propelled by an outboard motor, took us to the municipal landing in Avalon's harbor.

Shore Diving from Catalina

Ring Rock juts out of the water near shore; an iron ring is fixed into it. The rock is not particularly noteworthy, except that it marks the boundary of an underwater preserve off Catalina. Scuba diving and the taking of game are prohibited within the preserve. But snorkeling is permitted. A moderate surge was running. The entry over the rocks required some care and steady footwork, but was otherwise not particularly difficult. Once in the water, we descended while essentially swimming straight out from the shore and ultimately reached a maximum depth of sixty-five feet (nearly twenty meters). My partner showed me a brittle star; it was brown and black, and moved much faster than ordinary sea stars. Brittle stars resemble sea stars—common starfish—only in a very general way; both have five arms, unless the brittle star has lost one through deliberate autoamputation

119

—autotomy—which these animals use as a defensive maneuver. Sea star's thick arms taper smoothly from the center of the animal's body to blunt points, while the arms of the brittle star are pencil thin and wormlike. I held one of the brittle star's five arms; it promptly autoamputated the arm and escaped to grow a new limb. The severed limb crawled across my wrist as if it were looking to rejoin its fellows; I carefully grasped it between my thumb and forefinger and placed it on the seabed. It continued to writhe as I swam off.

We saw several kinds of goby. Gobies live on the bottom, often in small burrows made by other animals. We saw blackeyed gobies with their bulging black eyes; when young they are nearly transparent, whereas older individuals (at least they were larger and I therefore presumed them to be older) were an opaque milk white, still with bulging black eyes. Bluebanded gobies, *Lythrypnus dalli*, also called neon gobies, had stripes of electric blue along their sides, and they loitered among the spines of large Coronado sea urchins, *Centrostephanus coronatus*, safe from predators.

I was puzzled by many solid tubes, apparently of sand, that littered the bottom; they looked as if they had been extruded from a toothpaste tube. But if they were sand, what glue held them together, and what creature made them? Sensing that I was puzzled by what I saw, my partner picked up a nearby sea cucumber, pointed in turn to the sandy seabed, to the sea cucumber's mouth, its anus, and finally to the tubes of sand on the bottom. I understood at once: the sea cucumber, in this case *Parastichopus californicus*, eats sand and then defecates it in long tubes. I also guessed correctly that the explanation for this unusual feeding habit is that the sea cucumber lives on the organic particles that are mixed with the sand grains.

My partner cut open an urchin to feed the nearby fish; the bright orange garibaldis, *Hypsypops rubicunda*, proved the more aggressive (54), even bolder than much larger sheepheads, *Pimelometopon pulchrum*. The garibaldis had the field, and seemed to strike selectively at the yellow strap of my depth gauge. Reportedly, they will also strike the orange tape on the tips of divers' snorkels. Sheepheads, large fish of the wrasse family, pass through three stages in their life history: asexual juveniles, females, and finally males. They are protogynous hermaphrodites, meaning that they are capable of both sex roles at different times in their life cycle and are females first.

My dive partner chilled after forty minutes, so we exited the water over rocks covered with branched and fleshy green algae, *Codium fragile*, locally called sponge weed.

Catalina Kelp

Shore diving off Catalina can be excellent, but the best way to experience the fine groves of giant kelp that fringe the island is to board one of the charter diveboats that regularly visit these waters.

I awoke early in the little California coastal town of San Pedro. I did not shave or wash. I put on a bathing suit, pulled on a pair of coveralls, then stepped into my old loafers. I was ready.

We stopped for coffee-to-go on the brief drive to the pier where the big boat, *Westerly*, waited. Some divers were already on board. Others arrived in twos and threes over the next quarter of an hour, each carrying a large duffle bag and a scuba cylinder. After the last had arrived, the mate cast off and the captain worked her out of her berth.

We struck for Santa Catalina, about twenty-five miles from Los Angeles. The gray sea was smooth and a haze obscured the sun. Some of the divers were below deck, in bunks, finishing the sleep that had been so rudely interrupted by our early departure. Others were in the galley, pensive as they nursed their coffees. I was on the bridge, eager to learn from the captain where we were headed and what the diving would be like. Pilot whales cavorted several hundred yards off our starboard bow. After an hour or so, Santa Catalina lay ahead, hilltops hidden in mist. We dropped anchor hard by a rock rising up out of the sea, well away from the island. From a certain angle the rock looks like the prow of a ship. It is called Ship Rock and is home for a community of pelicans that have whitened the dark rock with their guano.

The captain's voice came over the public-address system. He described the underwater conditions at this dive site and cautioned us about local underwater hazards. We interrupted our preparations long enough to listen, then continued to pull on our wetsuits, don our inflatable life jackets, hoist our scuba cylinders onto our backs, step into our swim fins, and lower our masks over our faces as if they were the visors of knights in armor. Our final act on deck was to tug thick rubber mitts onto our hands. This is done last, since bare-handed dexterity is needed to adjust all the other pieces of equipment efficiently. Fully garbed, we shuffled aft in small groups, stepped off the transom and stood on a special diving platform hinged to the stern.

I poised briefly on the brink then let myself fall backward into the sea. The greenish blue water (unlike the gray-blue New England waters) rushed up and brought silence.

We glided along the submerged flanks of Ship Rock, about forty feet beneath the surface. My partner stopped frequently to take underwater photographs. I posed for her several times, each time pretending to fix my gaze on some photogenic animal or plant. A single tiny fringehead fish, *Neoclinus uninotatus*, darted out of a minute hole in the rock. It then backed in and disappeared. I watched the opening, expecting the fish to reappear. It did not. I saw the blur of a quick movement to my right, just at the edge of my field of vision. The fish had emerged at another point, several inches away from its first apparition. It popped back into the rock. On a hunch I waited a minute or so. I was right: the tiny fish poked its head from the rock again, this time in yet a third place. It dawned on me that this busy little creature did not live inside the rock but was exploiting the interstices left within some encrusting life form cemented to the rock. As far as I could see, it was fully protected from predators by means of its size; no other fish was small enough to pursue it into its hole (66). My partner spent several exposures trying to capture this fish on film. We finally broke free of its spell and moved on.

I spotted two dormant fish recumbent on a ledge and did a double take. They vaguely resembled sharks. But then I realized that they were harmless horn sharks.

In New England, sea anemones are a familiar sight. When these animals are active, their tentacles are extended and look somewhat like the flowers after which they are named. At other times, they curve their tentacles inward, toward their mouth opening, and draw the surrounding tissue over their mouth and tentacles. Thus retracted and enclosed, they look like freshly laundered socks that have been neatly rolled into balls. They are often yellow or brown. Here, I saw black oval lumps about six inches long and three inches high, each covered with what looked like black velvet. I immediately thought these were a variant of the sea anemones I had so often seen in New England.

SOUTHERN CALIFORNIA WATERS

I was wrong. This was an animal I had never seen before, the giant keyhole limpet, *Megathura crenulata*. It is a large single-shelled mollusk that creeps along the bottom, carrying its shell on its back. The shell is a flattened cone with a single hole piercing its apex. However, I did not see the shell, since it is normally completely enveloped by the animal's fleshy mantle, the black velvet covering I had first seen. The mantle does not close tightly over the top of the limpet but leaves the hole at its apex uncovered, and this aperture is the outlet for the strong current of water that the animal pumps through its body cavity to bring oxygen to its internal gills. I pulled a limpet off its rock, turned it over, and saw its yellow foot. Something made me glance over my shoulder. I saw my companions slowly moving away. I carefully placed the animal back, right side up, where I had found it. I kicked hard for a few fin strokes and rejoined the others.

I had hoped to dive in kelp, but we had seen none so far. I consulted my submersible air-pressure gauge and learned that I had already consumed half my air supply. I swam up to the other pair diving with us and gestured to the most experienced of the two. I made undulating vertical movements with my hands, trying to suggest long kelp plants swaying with the waves. Then I shrugged my shoulders, showing him my upturned palms. I hoped he would understand my message: "Where's the kelp?" He nodded and pointed ahead.

We four swam a little further along the rock, turning slowly to the right. Ahead was a plateau and on it a stand of giant kelp. I eagerly swam forward and entered the grove. The light took on a greenish cast and seemed peaceful. I looked closely at a kelp plant. It gripped the sea floor with a cluster of short radiating fingers, its holdfast. I raised my eyes to follow the lofty kelp

thallus to the surface. Leaflike blades arose in alternation from opposing sides of the stipe. Each blade was attached to its stipe by a short stem which had a central bulge, a float. I pressed one between my thumb and first finger and it yielded suddenly— hollow. The hundreds of floats carried by each plant buoy the fronds to the surface. There, the tangled stipes unravel and spread out to form a canopy of golden green. The fronds grow from their tips, new blades unfurling in a lovely repetitive pattern. The pattern expands from tiny bladelets at the tip downward to full-grown blades at the base.

Many animals and plants live in the shelter of the kelp forest. Deep-bodied, bright orange garibaldi played among the stipes, and schools of cigar-shaped senoritas moved fluidly through the grove.

Our guide beckoned. I swam to his side and he offered his palm. There, cradled in the hollow of his black rubber mitt, was a tiny, gorgeous creature. It was a nudibranch, the Spanish shawl, *Flabellina iodinea* (50–53). The animal's back was ridged with fleshy horns (cerata), which were light purple with orange tips. The colors seemed too intense to be emanating from so small an animal, perhaps only half an inch long. It lay in the hollow of my friend's hand, glowing like an ember.

I consulted the dial of my submersible air-pressure gauge. I should surface soon. I caught my companions' attention and beat my chest with a clenched fist, a signal meaning "I'm low on air." They nodded in acknowledgment and checked their own gauges. We headed up. We were still within our maximum safe time for this depth, but could not stay much longer without risk of contracting the bends—decompression sickness—upon or after surfacing. On the way up, two California sea lions, *Zalophus californianus*, dived down near us (49).

BENEATH COLD SEAS

We broke the surface and finned to the boat. My air supply was exhausted at this point, so I had switched to my snorkel; it was pleasant to pull moist sea air into my lungs after breathing the desert-dry air supplied by my scuba. I reached the stern platform and hauled myself onto it. I removed my scuba cylinder, passed it up, and climbed on deck.

I made two more dives that day. The captain obliged us by moving the boat twice, to two other good diving locations. I saw more kelp and more of the colorful inhabitants of the kelp forest. We also explored an ancient sea cave, now submerged under seventy feet of water. It was once at sea level, where centuries of crashing surf carved it from the rock. We did not enter its gaping maw because it was dark and we had no lamps.

I had one other remarkable experience in this very full day. We were swimming over an uneven rock bottom at about thirty feet when our guide stopped us near a crevice under a small boulder. He opened a plastic food container he had been carrying, removed a dead squid and dangled it over the hole, waving it slowly. A snout appeared, opened its toothy jaws and snapped the squid in two. Then it lunged forward to catch both severed morsels. More squid was offered. The snout took it, too. I soon realized that this was a moray eel. Our guide must have enjoyed the wide-eyed amazement of three divers from the East, for in a show of bravado he pulled off one of his thick mitts and held a piece of squid in his bare fingers. The moray came and took it politely, without nibbling the fingers of his benefactor. I was not in the least tempted to remove my mitt. The guide then caressed the eel's cheek. The eel responded by gently rubbing the diver's hand. (I do not recommend this potentially very dangerous procedure!)

After the last dive we spent another pleasant two hours motoring back to San Pedro. The sun had burned off the morning mist, and pilot whales again made long fluid leaps off our bows.

California's Northern Channel Islands

Approximately twenty-five miles due east of Catalina lies Santa Barbara Island, a small speck on the map, named by the explorer Sebastian Visciano in 1602 as he sailed past on Saint Barbara's day. Santa Barbara is open to the public for fishing, camping, day visits, and for diving, as is Anacapa. Although it is closer to the southern group of channel islands, Santa Barbara, together with the northern channel islands San Miguel and Anacapa, are collectively the Channel Islands National Park. Of these islands, San Miguel is the westernmost and is especially worth the effort to dive its waters and hike its hills, though a special permit is required to go ashore, and shore parties are limited to ten or under, in order to protect the island's unique environment.

The sizes of the northern channel islands are, from largest to smallest, Santa Cruz, Santa Rosa, San Nicholas, San Miguel, Santa Barbara, Anacapa (Anacapa is really three flat-topped rocks sticking up out of the sea). Both Santa Barbara and San Miguel are hilly islands clothed in grass and low scrub. But Santa Cruz and Santa Rosa are mountainous, with the usual scrub and grass but with trees as well. (The coreopsis on San Miguel are not really trees, but are related to sunflowers.) There is an olive orchard on Santa Cruz, and Santa Cruz and Santa Rosa both have cattle ranches. Santa Barbara is very much like San Miguel, while San Nicholas is barren.

I went on deck to find the large dive boat *Truth* underway in a moderate swell. We were headed for the waters off San Miguel. The sky was overcast and the deck was wet with a light but steady drizzle. I experienced a minor shock when I could not find my dive bag on the quarterdeck, where I had stowed it the night before. I thought it had fallen overboard in the night, but then I found it where one of the crew must have put it, amidships by the side of the deckhouse. It was precisely, but not I think intentionally, placed to take the full spray from the bows; the bag was not a waterproof one and it was soaked. Of course, water does no harm to diving gear, but the day promised to be gray and cold and I had hoped at least to have my equipment dry when I put it on.

I found no one in the ship's lounge, but was cheered by the sight and smell of a full pot of freshly brewed coffee. I took only half a cup, a limit I had resolved to observe on diving days. The reason for this was simple and compelling—I wanted to avoid having to remove my wetsuit between dives. Some cold-water divers do relieve themselves freely in their wetsuits while they are in the water, and they may enjoy the pleasant warmth that spreads over the belly and thighs. Yet this can easily become an unsavory habit if many dives are made each day and if several days of diving are performed consecutively. However, even the hardiest of cold-water divers would never do this in a drysuit, since there is normally no diluting saltwater in that garment.

By 6:10 A.M. some of the other passengers showed their faces in the lounge. One asked me, "How much longer?" "Don't know," I replied. One of them looked sick. At 6:30, the sun emerged from the fog. It had already risen above the horizon.

Though the sun was out the boat was pitching and rolling, if anything, worse than before. At 6:45, the sun disappeared again.

The first dive of the day was to forty feet (nearly twelve meters) for forty-five minutes. The second dive was deeper and therefore necessarily shorter in order to avoid decompression sickness; it was to ninety feet (about twenty-seven meters) for fourteen minutes with a five-minute stop at ten feet (three meters). This second dive was on a pinnacle of some kind.

On the first dive I saw a twenty-two-arm sunflower star, *Pycnopodia helianthoides* (45). I actually counted the arms, losing my place once and having to start over. This star was quite beautiful; its gray-brown body was covered with tufts of light blue and dark blue, and overall it had a white veil of tiny pincers, pedicellariae, having a protective function.

At first I thought there was poor underwater visibility; the "viz," to use divers' slang, was twenty feet or less. But then I saw that the water was filled with countless tiny larvae of some kind, each about one-quarter inch long.

The bottom was littered with bat stars, also called webbed sea stars, of at least five distinct colors: brown, gray, red, yellow, and tan. Some had textilelike markings, as if the patterns were of woven fabric (48, 75, 97). As for fish, there were good numbers of kelp bass and big sheepheads, as well as scores of milling blue rockfish, *Sebastes mystinus*, and schools of senoritas. I was not in the kelp yet. I surfaced to look for signs of the kelp mat; I saw it a few tens of yards away and took a compass bearing. I resubmerged to follow my compass and came to the edge of the grove. The light was dark brown, not greenish gold as it had been in the groves off San Clemente and Santa Catalina. The kelp seemed denser off San Miguel.

I entered the dark grove with some trepidation. The kelp

arched over me and lay flat on the top of the water forty feet above, nearly covering the surface. The grove was like a cave. Feeble sunbeams pierced the kelp mat and when I looked up the grove's ceiling looked like a thatched roof in poor repair. When I looked down I was startled to see what looked like a pair of large velvet footballs. I nudged one off its rock, and it tumbled in slow free fall to the bottom gravel. As it tumbled, I saw the muscular foot by which it had clung to the rock and which identified it as a mollusk. It also had a hole in the top of its hump. These could only be giant keyhole limpets.

I penetrated further into the grove and the brown light deepened. I followed a fissure in the bottom rock, knowing that such a natural landmark would make it easy for me to retrace my course and to find my way out. The bottom was exposed rock, with gravel filling the recesses. As I followed the crevice, the water shoaled. I looked up at the dense roof of kelp fronds, calculating that it would be easy to be entangled in them if I were to surface; I fumbled for the deflator tube of my floatation vest to expel every possible molecule of air, knowing that as I moved into shallower water any air left in my vest would expand with the decreasing water pressure, exert more buoyancy, and carry me upward. Thus a possible hazard of diving is to "fall" up in involuntary and uncontrolled ascent, accelerating as the gas in the floatation device or exposure suit expands as the diver rises.

Two seals appeared ahead. They were harbor seals, *Phoca vitulina*; their size, four to five feet long, their rounded bodies, and their short muzzles made identification easy. Not knowing if they resent human intruders, I was careful to move slowly while they were so near. They moved swiftly past, deeper into the grove. I then saw a bottom shark with a gently rounded midsection; it was a harmless swell shark, *Cephaloscyllium*

ventriosum, its body mottled with brown and spotted with white (69, 70). In looking around for other wildlife, I noted that there were no garibaldis. Nor had I seen any on the dives I had made off these northern channel islands; yet the garibaldi is ubiquitous in the kelp beds to the south, off Clemente and Catalina.

I saw two nudibranchs on this dive; both had purple bodies and brown hornlike cerata with white tips, and each had an orange blaze on its forehead. This was *Hermissenda crassicornis*, one of the species of nudibranchs known to eat corals, sea anemones, and hydroids, all of which possess stinging cells, nematocysts. The nematocysts remain undigested and pass through the predatory nudibranch to its cerata, where they serve as potent protection for their new host (101).

This was a wonderful dive. Many small sea anemones of several kinds dotted the rocks, and in a patch of sand between rocks I discovered what I suspected to be a tube anemone, which had laid sticky threads on the bottom like the spokes of a wheel, the tube anemone at the hub. When I disturbed the threads by touching them or by fanning the water above the animal, the threads began to move; the animal hauled them in slowly, taking several seconds to pull nearly their full length into its tube beneath the sand. The animal stopped short of pulling the threads entirely out of view, yet so short a length of them was left above the rim of the tube that they stood up stiffly; they had seemed limp before. The overall effect was of a quiver crammed with arrows. I was to notice these creatures many more times on these dives off the northern channel islands, always on soft, muddy bottoms.

The next dive was deep and brief, as deep dives must be. The water at this depth was colder and the light attenuated. I saw many kinds of sponges that I did not recognize, but I filed them

125

in my memory by naming them after familiar objects: ear (on a stalk), toothpaste, potato, ball. The sudden surprise of being deep when a shallow dive had been expected, as well as the giddiness that initially accompanies a rapid descent to below about sixty feet (eighteen meters) made my impression of the bottom life a kaleidoscopic one. I saw a nudibranch but did not think to note the kind; I also saw a blue and yellow top shell, a cone about half an inch at the base and about as high. Its color was a deep, electric blue with a gold stripe spiraled about it. It was the purple-ringed top snail, *Callistoma annulatum*. But I was mainly impressed by the variety and profusion of the sponges.

Several divers speared some colorful rockfish and took them up. On deck these fish were bright vermilion, but at ninety feet (twenty-seven meters) they had been a dingy brown. On this deep dive I saw fewer *Macrocystis pyrifera*, the most common species of giant kelp in these waters. I did see a purple, branching algae, as well as some big, flat laminarians. One diver brought up some bryozoans; they looked like algae, but they are in fact animals. They resembled potato chips, with one edge cemented to a rock, and were brittle, not fleshy like most algae.

After this deep dive the captain moved *Truth* closer to the rocks of Point Bennett, off the western tip of San Miguel. We saw many sea lions hauled out on the rocks. Sea lions have more elongated bodies than harbor seals do, and they grow bigger, the largest male sea lions reaching 600 pounds. When wet, these magnificent animals appear black, but they turn brown after their fur has dried as they bask on the rocks. Unlike harbor seals, sea lions have visible external ears, their hind flippers are adapted for locomotion on land, and they bark hoarsely. The sea lions shared the rocks with cormorants, long-necked diving birds whose guano has a very disagreeable odor; we could smell it quite well even

though we were anchored several hundred yards off the rocks.

I made two dives in the same place; the first was a skin dive. I ranged from the boat to within several tens of yards of the rocks, trying to get closer to the rookery. The water was thirty to forty feet deep and shoaled toward shore.

My objective was to get a closer look at the sea lions, both above and below water. I snorkeled in toward the rocks on which, it seemed, hundreds of the mammals were perched. The water surface bisected my faceplate, so I peered over the water as a professor I once had used to peer through the tops of his bifocals. I eyed the sea lions on the rocks and watched as a group of juveniles in the water moved into the territory of a big bull, who promptly slid into the water. He lunged and barked furiously at the boisterous trespassers, who moved on. As I approached the rocks, several sea lions leapt into the sea and treaded water, keeping their heads high to watch me. When I got too close, they dived and came to within ten or fifteen feet of me underwater. They swam in a swift circle about me, and several swerved toward me, then veered away. I knew of no documented cases of seals or sea lions ramming or biting humans in the water, but I kept my distance, just in case. Their wonderful speed and agility told of great body strength. I guessed that a blow from their flukes could be formidable.

I made five or six breath-hold dives to or near the bottom, at about thirty feet (nine meters). The seabed was paved with cobbles, and there were many sea urchins but few fish. Even after moving well away from the rocks, occasional solitary sea lions cruised past at blurring speed. Some seemed to alter their course to buzz me, or was this an illusion fueled by an overanxious imagination? Actually, I think they were only foraging.

I snorkeled back to *Truth*, boarded her and donned my scuba. During the scuba dive I thought I could hear the bark of sea lions underwater. I turned several sea urchins upside down and watched them right themselves. Like sea stars, they were fairly adept at this maneuver, and did it by moving their pincushion of spines in a coordinated fashion. I also saw a number of keyhole limpets on the rock wall that was the submerged foot of the sea cliff. Again, I noticed the absence of garibaldis. There were only occasional pillars of kelp, each comprising a dozen or more individual plants; at their common base was a dense mass of superimposed holdfasts. The stipes rose in a compact bundle to the surface, where they unfurled to form an overspreading canopy. The long stipes are often torn away from their holdfasts by the violence of the winter storms and are swept away. But the holdfasts endure to give rise to new blades the following spring. This may explain why some of the holdfasts are many inches thick, and among this gnarled and interlacing network a number of small crustaceans and mollusks take shelter. The stipes possess the bladders that buoy the plants up in the water, and each additional foot of stipe contributes more bladders and thus more buoyancy. I saw several examples in which this normally advantageous process—it provides light and water circulation—worked against their survival. The anchoring rock was too small, so that the combined buoyancy of the stipes had lifted it off the bottom. The plants were floating free, as plankton, moved by wind and current and probably doomed not to reproduce.

I picked up a sea cucumber, *Parastichopus parvimensis*, that was moving slowly across the bottom (99). It puffed itself up. I had noticed this behavior before. Is it a defensive maneuver that makes the sea cucumber too big to eat, or does it intimidate potential predators? I set the fat sea cucumber back on its track and drifted over red algae that looked like clusters of balloons; I had not seen such plants before. I encountered a cruising sea lion. We both saw each other simultaneously, and he turned sharply and came at me. I backed off, then he turned away. It is quite possible that he was not being at all aggressive, simply curious, and that I misinterpreted his actions. How many misunderstandings begin this way?

California Sand Bottoms

Sand bottoms have a subtle fascination. Many animals have found ways to conceal themselves, often in plain sight; on sand bottoms, things are often not what they seem.

Again, I was diving from *Truth* and again off the northern channel islands, but this time on a sand bottom with occasional outcrops of rock. It was here that I had an uplifting experience. In the bright but low-angled light the rocks cast deep shadows. There was a full palette of shades, and it was into this chiaroscuro world that I dived one late afternoon in the failing light. I felt a slight melancholy.

On the mostly sand bottom I saw a pair of swell sharks and scores of nudibranchs, especially the colorful Spanish shawls. Small fish came close; they seemed curious. But I was most impressed by the angel sharks, *Squatina californica*.

At first the angel sharks were difficult to recognize, since their flattened bodies were often completely covered by sand, but there usually remained a trace of their distinctive outline, like that of an angel with half-folded wings (65). This outline was often as sharp as if a knife had been thrust into the sand and run round the shark. Often, I would only see a fragment of this distinctive shape but could recognize it easily. An angel shark was concealed, except for one of its gills. I found and grabbed its tail. It levitated off the bottom and swam off. I also saw a flatfish with a dark, C-shaped mark near its head and a purple bull's eye near the base of its tail. This was a C-O turbot, *Pleuronicthys coenosus*, named for its markings. As I moved on, I rounded a low rock and glided over a hollow in the sand; I saw the outlines of two angels, one just overlapping the other. They seemed asleep, though I doubted they were. The tips of their tails were turned, so that their vertical fins lay flat on the sand. This shark, like so many other inhabitants of the sand surface, has adopted extreme flatness as its modus vivendi.

I knew that several divers had speared halibut on this bottom. Halibut, too, are inhabitants of sand bottoms, and like angel sharks are flat, though whereas the angel rests on its belly, the halibut lies on its side. One of the halibut's eyes would see only sand if it did not migrate when the fish is very young to the upturned side to join the other eye. If you look closely at a halibut's face—or a flounder's—it is easy to tell which eye started out there and which eye migrated from the other side. There are both right-eyed and left-eyed individuals.

The rugged cold-water divers that were aboard *Truth* brought a number of good-sized halibut aboard; these impressive fish weighed thirty pounds or more. They were California halibut, *Paralichthys californicus* and included both right-eyed and left-eyed specimens. Each successful hunter would wrestle his or her halibut up onto the deck. There would be brief excitement as the thirty pounds of nearly solid muscle exploded in a terminal paroxysm of fury and fright, slapping violently against the deck. The creatures' final struggle reminded me of the well-known line from Dylan Thomas's poem about not submitting easily to death: "Do not go gentle into that good night." A coup de grâce was administered efficiently with a rain of blows from stout clubs wielded by divers who rushed up to dispatch the beasts. Though this appeared brutal it was not, because it was quick. As the number of boated fish grew, the deck became flecked with the surrealistically bright crimson that only fresh blood can show. As the last of the divers boarded, the stern racks held a forest of spears. Each diver had brought at least two, for this was a group of avid hunters. San Miguel could be seen off the rail, wisps of pink sage on its green-brown hills.

The California halibut is a big fish, and can weigh more than seventy pounds. It is beautiful, even as it lies dying on the deck of a boat. Its sleek body is flat but with subtle fullness. This great and magnificent flatfish is but another example of the dictum that perfection is that from which nothing may be taken away.

A big halibut was dying on the deck. One diver did not see it lying there and stumbled against him. The creature's eye, silver and unblinking, could show no pain. The diver looked down and remarked with annoyance, "He's so ugly." The fish was soon dragged aft to the cleaning table, where it was quickly transfigured into a few pounds of filets.

I woke on a new day and went up on deck. The morning was foggy but with a promise of sun. All sins—at least those pertaining to decompression—are forgiven, since excess gas is assumed to have completely left a diver's body during a twelve-hour interval spent at the surface. Thus a new day can mean more diving. Fog shrouds Prince Island, a low three-humped rock off San Miguel. At exactly five minutes to seven, the sun broke through. At first the sun just limned the island; then, as the sky brightened, Prince Island went totally flat and became a cutout of black paper.

The first dive of the day was again on a sand bottom. I saw many sanddabs, small flatfish one-half to three inches long, with sand-colored bodies and a pattern that mimics the texture of sand so well that these small fish appear to have a dusting of sand grains on their sides. A swimming angel approached a spot on the bottom, circled once and then made a graceful landing. In the sand I saw many double siphons of some kind of mollusk, a bivalve like a New England razor clam. Possibly these were *Solen rosaceus*, the rosy razor clam of these California waters. When I touched the siphons they retracted well beneath the sand; and as they did so, the walls of the hole collapsed in an ever widening ring, leaving only a crater.

I saw a new outline in the sand; it was nearly circular. I deliberately irritated its owner by kicking water and sand at it. It rose up and shook the sand off its wings with indignity. It was a ray, a flattened fish related to sharks, and like them having a cartilaginous, not bony, skeleton. I guessed it was a Pacific electric ray, *Torpedo californica* (67). Its later aggressive behavior confirmed my tentative identification. It began to swim and I followed. After a short pursuit it became exasperated and turned to confront me. This aggressive behavior suggested that it had an offensive weapon. The electric ray's special threat is its ability to produce brief electric current. The source of this current is a battery of thousands of modified muscle cells, located in the lateral wings of the ray's compressed body. Ordinary muscle cells (as well as nerve cells) in all animals are able to produce a small voltage across their cell membranes, and this voltage is discharged, as an electric current, when the nerve or muscle cell

fires. I did not accept the proffered challenge: I swam on and left the ray in peace.

A cluster of tube anemones were cemented to a rock. I stroked the whiplike tentacles of one of these sedentary animals with my gloved hand; they were slow to retreat. Nearby, featherlike structures protruded above the sand; I touched one of them, too. The central spine was hard, though the other parts were soft and feathery. When touched, the feather was pulled smoothly beneath the sand. This animal is the sea pen, *Stylatula elongata*, and is related to the sea anemones.

I discovered a small octopus cowering under a rock. Its one visible eye was striking; it had a white iris and its pupil was a horizontal slit, reminding me very much of a scored aspirin tablet. I exposed the octopus by pulling its rock off the bottom. It was much smaller than I had thought when I had only seen its eye and part of its head—only about four inches in diameter. It crouched in the cavity left by the rock and then folded its tentacles and tried to jet away. It was easy to capture. I put it on my wrist, where it clung to the nylon fabric covering my rubber suit. It curled the tips of its tentacles in tight little spirals. Several big sheepheads eyed it and circled us; the octopus was exposed and vulnerable, and I immediately felt protective. I held it in one hand; it wanted to escape. Did it sense the danger? I tried to restrain it without injuring its soft body. With my free hand, I worked its dislodged rock back into its original position. Then, I guided it into its old hole between the rock and the sand. It hesitated; I gave it a gentle shove, and it disappeared under the rock. It pulled back deeper into the shadows this time, and I could no longer see its eyes. I moved on.

On the next dive I searched for my first halibut, but found none, although several divers bagged thirty-pounders. Sand dollars were clustered in patches on sand that was deeply sculptured with wave ripples. A fairly strong underwater current was running. Many large sea hares, *Aplysia californica*, ten to twelve inches long, lay on the sand, nearly always on or near some kind of algae. The largest of them were as big as loaves of bread. I saw some big kelp I had not seen before; it had a straplike longitudinal member with thinner filaments fringing its edges, feather-boa kelp, *Egregia laevigata*. Another kelp resembled fir boughs, with a long central strand and with rows of tiny bladders on some of the lateral branchlets; these algae were ten to thirty feet long and many were lying free on the sand, as if they had come to rest after being torn loose in a storm. This was bladder chain kelp, *Cytoseira osmundacea*. However, the sea hares seemed the most abundant in the vicinity of small translucent, green algae. True to its name, the sea hare is a vegetarian.

The following three dives were on mostly sand bottoms and all with the same objective; to find the great flatfish, the California halibut. On the first dive I swam across a sand bottom with close-spaced ripples and sand dollars stuck into the sand on their edges, like cookies in a custard. But no halibut. Dive two was in large kelp of mixed species, and I saw the usual fish and other animals one sees in a kelp grove in these waters. But still, no big flatfish.

On my third dive I glided over the bottom toward shore across long, low sand ripples. There were no sand dollars. I was cruising through only moderate underwater visibility, twenty feet or so, nearly grazing the peaks of the sand ripples on each downstroke of my kick. There it was. Motionless. I slowed, stopped, and watched. I could see no movement of its gills. It was a sleek gray-green hump on the sand, and it was uncovered—no sand on it—a sign that it may not have been there long. It was

big, at least a thirty-pounder. I looked at its head; it was still except for its eyes, little periscopes that rotated to scan for danger. I lay next to it. We eyed each other and I crept closer to where I could have touched it. But I had pressed my luck too far; it sprang off the sand with no apparent effort and was in undulating flight. I took off in pursuit. It quickened its stroke and led me in a half circle, then veered off into the underwater haze. It was much too fast to follow. I checked my compass and made off in the approximate direction of *Truth*.

As I neared the big boat I could hear the hum of her diesels. I sensed a shadow overhead and looked up to see her hull in dark silhouette. I ascended at the stern, hauled out onto the ramp, and stood up. One of the divemasters, who was checking off divers as they boarded, asked me the customary question, "Wh'dya get?" "A big one," I replied. A knowing, satisfied look came over his face, but quickly turned to mild puzzlement as he looked me over but saw no sign of a fish. "Where is he?" I raised my hand to the side of my head and pointed a forefinger at my temple "Up here," I said. There is more than one way to capture a fish, and as a writer I had captured him in my own way, just as surely as if I had thrust a spear point into a smooth, brown flank. But my questioner was doubly puzzled now. Was I playing some obscure joke on him? Was I slightly crazy? He ultimately mastered his puzzlement simply by acting as if I had never said a thing, as if it had never happened. He looked down at his roll of names, checked off the diver boarding just behind me, then turned to him and said, "Wha'dya get?"

Rocky Bottoms

California rock bottoms are especially rich in life. Groves of giant kelp inhabit rock bottoms, since their holdfasts must have a hard surface for anchorage. But there is a complete community of organisms in California's rocky bottoms; the abalone is perhaps the most famous of these rock dwellers.

There are California rock bottoms where the seabed is nearly paved with abalone. I dived at one of those places. Many of the abalone were more than seven inches across their major diameter. Again, I tested their strength. I found I could not pull even a four-incher off its rock, with a steady hard pull. In one case I only succeeded in wrenching a small boulder out of the bottom mud. After the dive, several divers told me that they had in the past pulled abalones off the rock—legal ones, they asserted—by hand, without benefit of the special tool usually used for this purpose, an abalone iron. The secret, they said, was to surprise the animal while it was standing up, its shell high off the rock and its foot extended. Once alerted the abalone hunkers down, clinging tightly to its rock and hardly leaving enough space under the rim of its shell to slip gloved fingertips. I knew that my experimenting was potentially dangerous. I could feel the strength of the animal as it tightened down on my fingers. I had heard stories of hapless divers, their hands powerfully clamped to a rock by a giant abalone, running out of air and being helpless to turn on their reserve air supply. With these probably apocryphal tales in mind, I was never tempted to thrust more than the very tips of my fingers under an abalone's shell. If the animal began to clamp down too hard, it was my plan to slip my hands out of my gloves and to leave the gloves pinned to the rock.

I found a small kelp grove near the end of this dive. My eyes were drawn upward by the stately columns; I looked up from the bottom and once again saw how lovely the brown, leafy canopy is. It was there that I saw two garibaldis, my first sighting of these fish on any dive north of Santa Catalina. I also saw the familiar sheepheads and calicos. I ascended to a hole in the kelp mat and at the surface I took a bearing on the boat. I redescended and followed my compass until I could hear and feel the throbbing of *Truth*'s diesel generator. I looked up and saw her silhouette. As I ascended I was at the center of a contracting circle of bright light on the water's undersurface, an optical phenomenon caused by reflection. I surfaced at *Truth*'s stern ramp and bellied onto it.

The next dive of the day was also on rocky bottom, a pinnacle off Carrington Point on Santa Rosa Island. There was a profusion of sea stars on the seabed; sea stars and rocky bottoms seem to go together. There was also a soft carpet of sponges on many of the rocks.

A California cone shell, *Conus californicus*, was gliding slowly over a naked rock, with its white proboscis extended. Cone shells are poisonous, though unlike some of their relatives in other parts of the world, their toxin is not thought to be life-threatening to humans; but they should be handled with care or—better still—not at all. The cone shell's tonguelike radula has been modified to form a harpoon, a stinging apparatus that is used to inject poison into its prey. The same rock supported an enormous sea anemone, approximately twelve inches in diameter at its tentacle tips; it had white tentacles and a brown center, a species of the genus *Tealia* (82). My rubber mitts dragged slightly as I brushed its tentacles, caught and stung by the

131

animal's barbed nematocysts.

A confusion of brittle stars was piled in a hole beneath an overhanging rock. Their pill-like bodies and stringy arms identified them as the smooth brittle star, *Ophioplocus esmarki*.

Whole rock surfaces were covered with strawberry club anemones. They do, from a distance, show red and white in about the correct proportions to suggest strawberries. From these rocks also sprouted pale pink moonglow anemones, *Anthopleura artemesia*, as well as the beautiful sand rose anemone, *Tealia columbiana*, its thick red stalk ringed with white spots and surmounted by a diadem of white tentacles. And among the sea anemones were fluffy clusters of feathers, the exposed gills and feeding apparatus of feather-duster worms, *Eudistylia polymorpha*.

Several large sunflower stars lay on clear surfaces of rock; they had a regular pattern of fine blue and white tufts on their backs, and a fine branching network of orange ran just on or under the skin, among the tufts. Moon sponges, pale and cratered—like the surface of the moon—thickly padded some of the rocks, many with tube worms inhabiting the holes. Jointed purple algae, the beautiful and abundant calcareous *Corallina*, filled in the gaps left by sponges and sea anemones. In midwater, a school of strange and pencillike tube snouts, *Aulorhynchus flavidus*, drifted by.

I looked down at the holdfast mound of a kelp cluster; within the interstices formed by the many layers of overlapping and intersecting tendrils was a small world of worms, tiny starfish, fish that looked like tiny kelp blades, and many other animals I could not identify. As I finned forward I came to a rock wall.

Large crabs shared these rocks with the other inhabitants. There were decorator crabs that cover themselves with a living disguise of sponges, hydroids, and tunicates. Blue Dungeness crabs, *Cancer magister*, appeared in small groups, a large individual usually associated with one or more smaller individuals in a single crevice. Large sheepheads and several kinds of rockfish completed the scene.

I enjoyed ascending alongside a column of giant kelp plants. The low afternoon sun filtering through their blades cast a warm, brown light.

The captain moved *Truth*, and I made yet another dive. The bottom was rock and deeper than I had dived for several days. I saw some giant kelp, but it was sparse here. There were also large moon sponges, pale and cratered. A heavy surge even made itself felt at sixty feet. A pair of sea lions flew by and ignored me. I sneezed underwater for the first time in my life, two volleys of four sneezes each. My mask stayed put. I noted with interest that each sneeze was preceded by a very hard, long inhalation that seemed involuntary. I gripped my mouthpiece tightly. After the sneezing subsided I removed my mask to blow my nose.

Several big lingcod lay on the bottom rock. They are long tapering fish and are not true codfish. They have enormous mouths that are full of teeth. Once cleaned and filleted, their flesh is the color of pistachio ice cream.

I soon found myself in dense kelp forest. I noticed kelp blades had round floats near each plant's holdfast, and that the floats gradually became egg-shaped near the apex of the plant. This was something I had not noticed before. As in many of the other kelp groves I had seen, the plants were grouped together in columnar bundles of six to ten plants each, and these columns were spaced ten to twenty feet apart in all directions to the limit of my underwater visibility. Again, large lingcod lay on the bottom, and sheepheads moved among the kelp. I also saw several vermilion rockfish, dusky brown at sixty feet and orange at thirty feet. They showed their bright vermilion only at the surface.

I moved to shallower waters. At thirty feet I saw several types of red algae. One was branched with thumblike bladelets—*Rhodoglossum* or red tongue, closely related to the Irish moss of New England. A second kind of algae had large, flat oval blades with bumps; these were a species of *Gigartina*. Black keyhole limpets—and lighter, speckled ones too—worked their way across some of the rocks. I saw a green algae I termed head of hair. It was *Chaetomorpha aerea*, and it moved like a woman's long hair as she throws her head back in a breeze.

A California sea lion surprised me at thirty feet, coming up close and peering into my faceplate. I had not seen it approach, so I was startled when I saw it staring at me. It circled once and swam off; I think it simply wanted to play. Other divers have told me that they have been similarly approached, and that sometimes the sea lion blew a burst of bubbles in imitation, at least one diver thought, of the diver's own exhalation.

Above thirty feet I saw the pink jointed algae, *Corallina*, and green algae I termed gherkin, because their fat tubular shape reminded me of little pickles. I learned that it was *Codium fragile*.

I found the anchor line, ascended it and swam directly under *Truth*, hugging its hull. As I neared the stern I saw a cloud of fine bubbles. The propellers, to which I had strayed dangerously close, were cutting the water just a few feet from my face. My only thought was to dive. The dark hull glided over me and the props noiselessly churned the water just above. The lesson is obvious, and was not lost on me.

I popped to the surface well to one side of the boat in an area clear of kelp mat. Although from below water the tops of the giant kelp form an appealing translucent canopy that filters the sunlight and renders it a golden brown, at the surface the same

kelp tops form a seemingly impenetrable mat. Actually, it is fairly easy to drop through vertically, but it is extremely detentive if any diver attempts to move through it horizontally; many divers have tired themselves struggling against it. The trick is to swim underwater, where the kelp is held in taut columns by their flotation bladders. At the end of a dive it is best to ascend and penetrate the canopy very close to the side of the boat; or, if the boat is still distant, to take its bearing with a compass, estimate the distance to it, and then resubmerge and swim underwater well beneath the canopy to as nearly beneath the boat as one can navigate. With practice one can become surprisingly adept at finding the boat while underwater; swimming perpendicularly to the anchor line helps, especially when the skipper has payed out lots of scope; then the anchor line rests almost flat on the bottom near the anchor. This long underwater marker greatly increases one's chances of finding the boat and it will lead directly to the bow. However, the divers' boarding ramp is at the stern. Do not be tempted, as I was, to swim just under the hull to reach it.

I was now safely on deck, but I saw a diver in distress. He surfaced in the midst of the kelp mat far from the boat and was out of air to boot, so he could not redescend and approach *Truth* underwater. He compounded his chain of errors yet further by struggling against the clinging brown tendrils. He cried out in panic and a crew member accurately threw him an inner tube tethered to a long line and towed him over the tops of the kelp to the side of the boat. A second out-of-air diver surfaced in the midst of the dense mat. A strong breeze began to swing the big boat on its anchor line toward the diver. *Truth*'s many tons would have slid smoothly over him, submerging him. In his exhausted state the kelp kept him from escaping. Those of us who saw this happen endured a moment or two of nearly unbearable suspense. The divemaster jumped in to pull him clear of the hull

133

SOUTHERN CALIFORNIA WATERS

just in time.

I shared the sunset with the other divers on board. We were anchored off San Miguel Island and its satellite, Prince Island. The islands were both stark silhouettes against a reddening sky. The steel blue water was streaked by a stiffening breeze, and kelp floated at the surface in brown masses. The mixture of colors —black, red, blue, and brown—created a wild and beautiful effect.

Ashore on San Miguel

I woke at 5:45 A.M. and went on deck. At 6:30 the crew moved the boat into Cuyler Harbor, San Miguel Island. A small party was to go ashore. I took this opportunity, knowing it would be unlike any other experience I could have on the trip and that I would probably never have such a chance again. Six of us stepped down into an inflatable boat, powered by an outboard. We motored to the island, where we beached the boat and pulled it high up on the berm, out of reach of the tide. We felt like castaways as we watched *Truth* move out of the harbor for a day of diving. We trudged off down the beach toward a gap in the hills named Niedever Canyon. We followed this rising gully to the hut of the resident United States park rangers, volunteers who were college students in real life. The plant life on the island was fascinating and probably the result of drought and wind, the two prevailing conditions of the island's climate. Some of the plants were fleshy succulents: ice plant and dudlea. However, another plant was the most prominent feature of this side of the island: it grew in a stunted forest, two to four feet tall, of strange and seemingly half-dead dwarf trees. They were faintly green at the tops of their trunks and at the tips of some of their major branches. These plants looked woody but were flexible, so I knew they were alive; they were giant coreopsis. There was also flowering buckwheat, which formed a pink veil on these brown hills.

Our little expedition, led by the two rangers, hiked to a monument erected to the explorer Cabrillo. The monument, a simple stone cross, was inscribed:

João Rodrigues Cabrll'ho
Portuguese Navigator, Discoverer of California 1542
Isle of Burial 1543

The monument stood at the edge of a high meadow that commanded a view of the sea. Scattered in the grass were California poppies, their yellow corollas punctuated with orange centers.

We slowly hiked to the highest point on the island, San Miguel Peak, about eight hundred feet above sea level. Other plants caught my eye and were identified for me by the rangers: yellow chicory, golden bush, and coyotillo (also called coyote). Golden bush and coyotillo are both low shrubs. I thought I could just see Point Bennett due west, and as I looked down the dry hills of burnt brown to the sea, and then across the sea itself to the horizon, the sea and sky became one—a glimpse of infinity.

We returned the way we had come, over the peak, across the meadows, down dry gullies lined with low coreopsis. The strange landscape had an extraterrestrial appearance. We debouched from the canyon and were walking on the beach just as *Truth* came into view.

The final dive of the day was made at dusk. I entered the water at the bow, making a long leap. I swam to the anchor line and followed it down. Though the sun had just set, there was plenty of light underwater, especially after my eyes had dark-adapted for a few minutes. Everything seemed still, quiet. Perhaps my own impression of calm was conditioned by the time, the peaceful hour of dusk. At mid-depth I encountered the biggest school of fish I had seen so far, but its individual members were not moving in the frantic, coordinated way that makes a large school appear a streaming, silver fluid. Instead, they seemed to move in slow motion, as if I were dreaming them; and they all seemed to be looking up toward the boat. Were they expecting something? I knew that some of the other divers had called it a day, shucked their rubber suits for their warm clothes and had become pensive fishermen with rods and reels at *Truth*'s rail. Perhaps the fishing lures dancing well above my head held the attention of these thousands of pairs of eyes. As I passed through the school and approached the bottom at fifty feet I found a rocky reef peopled by rockfishes of various kinds and enormous starfish. There were also very large sea anemones, at least twelve inches across their tentacles, whose centers looked like brown puddings.

That evening, a *Truth* crew member showed me the machinery below deck. The engine room was dominated by two spotless twelve-cylinder diesel engines, their green-painted cast-iron blocks and gleaming chromed valve covers showing that they had been wiped regularly. "Dirty engines are hard to monitor for oil leaks," my guide said. Aft of the two propulsion diesels was a six-cylinder diesel-generator set, which provides *Truth*'s electrical power. Aft of the engine room is another compartment in which the boat's brace of large high-pressure air compressors stand; they provide the vital breathing air for the divers' scubas. This pair of diesel-driven compressors would be the envy of any diveshop ashore. Forward of the compressors were tanks to hold one thousand gallons of freshwater.

I later redescended into the engine room to hang my wet-suit to dry in its heat. I again experienced the magnificent music that those mighty engines make—the orderly roar of twenty-four pistons slamming through their cylinders.

On the run back to Santa Barbara we saw several blue sharks basking at the surface; they had long sleek bodies and we found them over very deep bottoms. We also passed a troop of cavorting sea lions porpoising in the water. The captain let me take the helm for thirty minutes or so and I found it difficult to keep *Truth* on a straight course, although she had power-assisted steering. The problem was that when I corrected her course, she would turn farther than I had intended; then I would correct in the opposite direction and she would go too far the other way. This dithering went on for some time. I looked back at our wake, which extended for several hundred yards behind us; it was a nearly symmetrical sinusoidal track. When we docked at Santa Barbara all the other divers disembarked, and those I had come to know made a point of saying good-bye. We then moved *Truth* to the fuel dock, where she took a long and costly drink of fuel. As I thought back on my week aboard this magnificent dive boat, John Keats's lines came to me:

> "*Beauty is truth, truth beauty,—that is all*
> *Ye know on earth, and all ye need to know.*"

135

SOUTHERN CALIFORNIA WATERS

4 / NORTH OF CONCEPTION

Monterey Waters

THE WATERS of the North Pacific Current move west until they reach the North American continent at a level that falls variably between Oregon and Washington, depending on the season and other factors. Here, at this continental barrier, the westward flow diverges into the north-moving Alaska current and the south-setting California current. As a result of this two-headed movement of cool water, the coastal waters of North America from the Gulf of Alaska to Point Conception—about fifty miles east of Santa Barbara, California—are remarkably uniform in temperature.

To the south of Point Conception, the coastline tends sharply to the east, away from the California Current. Thus, Point Conception divides two regions of the temperate north Pacific: the warm-temperate province to the south from the cold-temperate province to the north (though both regions are cold by human standards). A few animal and plant species are found in both provinces, though many occur predominantly either to the south or to the north of the point. However, there is no sharp dividing line, and in the waters surrounding Point Conception there is an area of overlap that extends at least as far north as the Monterey Peninsula. In this area a rich variety of both warm-temperate and cold-temperate marine life can be found.

Thus, diving in the Monterey area, approximately 150 miles north of Point Conception, is quite special. The great

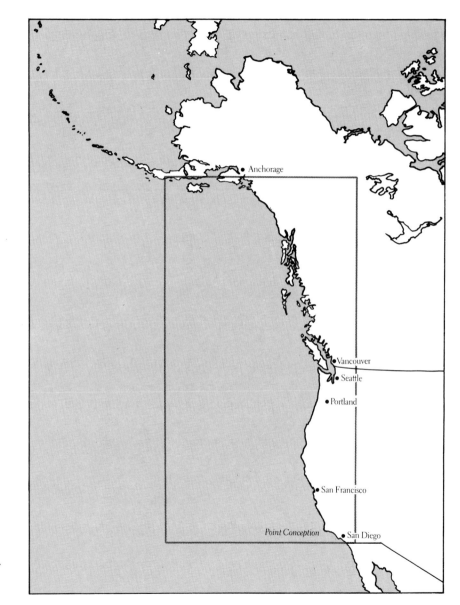

Figure 4.1

Along the west coast of North America, a vast cold-water region stretches from Alaska to San Diego.

NORTH OF CONCEPTION

Monterey and Carmel deep-water canyons are near-shore reservoirs of deep ocean water, laden with nutrients. Upwelling elevates this water to the surface to support a rich growth of organisms in the upper sunlit layers where photosynthesis can occur. The variety of marine life, the reservoirs of fertile deep waters close to the shore, and upwelling combine to make diving here particularly rewarding.

Point Lobos State Reserve, a few miles to the south of Monterey, is 1,250 acres of nearly unspoiled coast; it resembles what this part of the California coast must have looked like before it was intensely developed and much of its great natural beauty was lost.

We arrived at the entrance of the reserve early, knowing that the rangers closed the roadway when the allowed number of cars had passed through the gates. We were admitted precisely at 9:00 A.M. The rangers were courteous, but were quite thorough in checking our diving credentials and our safety equipment. They also informed us of the rules for divers: we were only permitted to dive in two adjacent coves, Whaler's and Bluefish, and we could enter and leave the water only at Whaler's. We were then waved through the gate into the park. We drove the short distance to Whaler's Cove, where we parked. We looked out over the cove and its thick kelp mat. Cannery Point bounded the cove to the west, Coal Chute Point to the east. We then hiked west, to the top of Viscaino Hill, which commanded a view of Bluefish Cove. The landscape was a synthesis of craggy rocks and clear blue water. The vegetation included the fleshy ice plant, wind-shaped Monterey cypress, and poison oak. The latter grew almost everywhere and helped to hold against erosion the soil that clung to the shallow recesses in the rocks.

We entered the water in Whaler's Cove and crawled over the kelp instead of swimming beneath it, in order to conserve our scuba air for use beneath the fine water of Bluefish Cove. It was a long, tiring swim. Finally we descended.

We reached a maximum depth of seventy-five feet (twenty-three meters). My first impression was of large sea anemones with bases like the trunks of small trees, rings of white tentacles, and yellow centers. The circles of tentacles were twelve to sixteen inches in diameter. Smaller anemones, six to eight inches in diameter, had pea-green tentacles. Yet another anemone had a white corona about ten inches in diameter on a slender yellow-green stalk at least eighteen inches long. A harbor seal adopted us on the way out and stayed with us for most of the dive. A ray appeared, possibly a stingray. Our companion seal nudged it as if to shoo it away.

In a mixed stand of giant kelp and bladder chain, a pair of nudibranchs grazed on the bottom; they were yellow with brown freckles, false sea lemons. There were also giant gumboot chitons, *Cryptochiton stelleri*, and hermit crabs in turban shells. Big lingcod blended well with the seabed; one was at least thirty inches long. A cabezon, *Scorpaenichthys marmoratus*, lay under a rock. It is an ugly, mottled fish related to the sculpin. A beautiful rainbow surfperch swam by in midwater, with its horizontal bands of orange and greenish blue. The harbor seal continued to dog us (his face even looked like that of a friendly mutt). Ahead were rock walls covered with sponges and sea anemones. Shallower rocks were hung with seaweed in striking variety: red wing-nut algae, *Bossiella* (102); sea palm, *Postelsia*, supporting a crown of fronds; and head-of-hair, streaming with the surge.

On our way back we headed east to Coal Chute Point, where we found a cave—an open tube—among tumbled rocks. A moderate swell moved overhead, and each wave made a smoke of bubbles as water was pumped through the opening in the rocks. We were very low on air. I rode through several times while using my reserve air, a bad practice. But it was exhilarating to be blown through the slick-sided tube by the heaving surge overhead. We finally surfaced, both low on air; in fact, I had virtually none. We had hoped to snorkel across Whaler's Cove to where we had entered, but again we found the kelp so thick that we had to crawl over it.

We returned to Point Lobos yet another day. This time we trailed a rubber boat. We paid admission to the reserve only for the purpose of using the boat-launching ramp at Whaler's Cove. We intended to go beyond the waters of the reserve for our diving. We motored cautiously through the kelp mat to the mouth of the cove, vigilant for divers' bubbles. Several little doglike heads bobbed ahead—sea otters—but they ducked under before we got close enough for a good look. At the mouth of the cove we were free of the kelp mat, which had attenuated the waves entering the cove; we now felt the full swell. My host showed no mercy: he gunned the outboard engine and we flew over the water. I sat sidesaddle on the boat's starboard sponson and my knuckles whitened around a rope. There were no thwarts to sit on. The boat did extremely well on the surface chop—better than I. The shore streaked by. The coastal range was shrouded in fog, but the dark green of the wind-shaped cypresses shone through.

We hove to off Yankee Point, south of Point Lobos. We anchored the boat by tying its mooring painter to a good-sized bundle of giant kelp stipes, just as sleeping sea otters, *Enhydra lutris*, moor themselves by wrapping a few turns of stipes about their bodies (57). Each of us attached his scuba to a lanyard and then threw it overboard. We then rolled into the water and donned our scubas in the water, a procedure made necessary by the impossibility of standing up safely in the small boat. We descended against a current that swept the tall kelp into a steep angle. We pulled ourselves down hand over hand on stout bundles of stipes. This was strenuous work; my host was in superb physical condition, but I was not.

We reached a maximum depth of 110 feet (33 meters), and the kelp extended from at least this depth to the surface. Sheer outcrops of rock were softened, colored, and brought to life by lush colonies of strawberry anemones. I also saw large solitary anemones as well; some on stout trunks, others on slender stalks. Big lingcod and numerous colorful rockfish of several species inhabited the grove. It was a beautiful vista; we only had 30- to 40-foot (9- to 12-meter) visibility, which I enjoyed, but this dive would have been truly spectacular with the 50- to 100-foot (15- to 30-meter) visibility that my host said prevails here in the best of times.

We moved the boat for a second dive over a rock pinnacle that rose up off the ocean floor south of Lobos Rocks. Again, there were impressive vertical rock faces covered with yellow cup coral and strawberry anemones. In deep water we saw the giant acorn barnacle, *Balanus nubilus*, with its distinctive paired beaks (79). I had thought that barnacles only lived in the intertidal zone; but here they were at nearly eighty feet. Large sea anemones abounded here too. These had fluffy white crowns of tentacles and stood on slender brown stalks. Striped convict fish, kelp greenling, lingcod, blue rockfish, black-and-yellow rockfish, china rockfish, and calico bass shared this place. Several small varieties of kelp, including sea palm, grew on the sides and tops of rock outcrops, and large *Macrocystis* bundles rose from the flats in between; their stipes were canted to an oblique angle due to the powerful current. Purple-ringed top shells grazed on branching hydrocorals. Giant chitons glided slowly over the rocks; each chiton had a warty back and a fleshy yellow foot. A huge, many-armed sunflower star lay curved in a recess, and a sea star with slender orange arms—a blood star, *Henricia leviuscula*—rested nearby. There was a fair amount of bull kelp, which I called minefield kelp because each plant had a large spiked bladder tethered to a long cordlike stipe. On the bottom, large turban shells glided over the bare rock, and giant barnacles grew in encrusting sponges. One very large, dense colony of hydrocorals was especially striking (73), and blue-banded top shells were grazing on many of their stony branches.

We moved the boat again to dive in a shallow kelp bed half a mile offshore from Soberanes Point. Here the bottom fell away in cliffs, and was littered with boulders and rubble. A well-dressed decorator crab had an orange sponge on his noselike rostrum (58, 85). I wondered what happened to his decorations when he molts; I learned later that he molts under cover of night and quickly decorates his new carapace. I got a closer look at a kelp greenling, *Hexagrammos decagrammus*, a beautiful brown fish with small blue spots near its head; its pectoral and dorsal fins were marked with black and gray. The kelp canopy was the thickest we had dived under. The light was deep brown, and it quickly became darker as we dived deeper into the grove. A bright orange sea star shouted out its vivid color as I swam by (86).

As we returned to Whaler's Cove, we saw sea lions in the water. Two species are found here: California sea lions and the larger Steller's sea lions. Near Seal Rocks they treaded water to watch us. A sea otter raised its head many yards away and I could

just make out its fat, furry body. As it dived, its little feet went up in the air. I had seen others yesterday, lying on their backs in the kelp mat, but could not get close; they dived too quickly. We rounded the rocks off Cannery Point, slowed as we entered the mouth of Whaler's Cove, and motored through the kelp to the boat ramp.

Big Sur

Big Sur is a stretch of rugged coast to the south of Monterey. The sea here is rich in underwater wildlife, but the cliffs that make this area an inexhaustible subject for romantic picture postcards make it difficult for divers to enter the water safely. There are, however, a few stretches of this shore where diving is possible.

We pulled onto a dirt road that was nearly hidden from the highway. We parked and got out to walk over a meadow of flowering ice plants and stood on a cliff overlooking the sea. The surf was up. My guide asked me a few questions about my surf-diving experience; he may have had some anxiety about taking me in the white water without having first-hand knowledge of my diving ability. I must have satisfied him, for we returned to the car, unpacked our equipment, and then trudged back to the cliff with all our equipment on our backs or in our hands. We worked our way down from the brow of the cliff to just a few feet from where big green breakers were rolling in. We watched wave sets, groups of three or more waves that are lower or higher than those preceding or following them. There was a strong surge due to the three- to six-foot swells; most were around five feet. There was a large rock just a few feet offshore from the place where we planned to enter the water. As the swells humped up, water surged even higher in the ten-foot gap between the rock and the shore. We watched this rise and fall for several seconds before we had each calculated our best entry point and course to follow out to deeper water. A set of large swells thundered in. I waited for them to dissipate, then entered quickly.

I snorkeled hard against the surge in the narrow gap, avoiding the white water near the rock. Once clear of the gap, I turned away from shore and snorkeled out to the edge of a kelp mat, fifty yards offshore. At the edge of what promised to be a dense grove of giant kelp, my partner and I exchanged the usual last-minute formalities. Then we dived beneath the dense canopy, making an oblique descent to the bottom. The grove was very dense indeed, and there was a tenebrous light. *Macrocystis* was not the only giant kelp here; bull kelp also sent up its slender cordlike stipes that thickened as they rose, some achieving the diameter of a man's wrist just below a nearly spherical bladder; each bladder was surmounted by a crest of flat blades. We moved out along the gently sloping bottom to a depth of sixty feet, where the glade was very dark. Even at that depth I could feel the heavy surge from the swells rolling over the surface. In the crepuscular gloom, my partner's exhaled bubbles were bright little explosions of light. I saw several medium-sized ling, and glimpsed a ghostly moon jelly, *Aurelia aurita*, making slow pulsatile progress among the kelp bundles. I swam alongside it and saw through its translucent bell the crossed figure eights formed by its bright orange gonads. The margin of the bell rhythmically contracted and relaxed to row the animal through the dark, liquid stillness. I also saw sponges, nudibranchs, bryozoans, hydrocorals, and sea palms.

As we reached a depth of ninety feet, the rocky bottom gave way to a plain of coarse gravel. Interspersed on the gravel plain were boulders and patches of sand. Even here I felt the rhythmic tug of the swell overhead. As our bottom time approached twenty-five minutes, I got my partner's attention and signaled that we should turn back; we have only thirty minutes of total bottom time on any single dive whose maximum depth is ninety feet. We consulted our compasses and set a course that would take us back to shore through the kelp.

It was much easier to make forward progress near the bottom where the *Macrocystis* are tightly fasciculated columns, but new to me were the cords of the bull kelp, some as slender as my little finger, but tough and sinewy; they seemed to be everywhere and were not confined in bundles like *Macrocystis*. Bull kelp strands found their way behind the protruding valve on my scuba, snagging me until I could reach back and disengage the kelp. Strands also caught the spare breathing mouthpiece hanging at my side, and other strands hooked onto my snorkel and nearly pulled my mask off my face. Several times I found that I was swimming without moving forward, only to discover that some part of my equipment was caught. As I released each tangled strand, it straightened slowly as its apical float rose and pulled the strand taut.

My partner did an excellent job of navigating. He brought us near shore, very close to where we had entered the water. Again, I watched the pattern of waves rolling shoreward and saw a set of smaller ones. I rode them into the narrow chute that I had followed out to sea at the beginning of the dive. The waves neatly deposited me on the shore. Knowing what was going to happen next, I hung on to the rocks with all my strength and found footholds as well. The same uprush that had carried me in was now, driven by gravity, streaming backwards in retreat and nearly

141

swept me with it. The rocks around me were a hundred waterfalls as tons of water drained away. Shortly the rocks were simply wet, and I scrambled awkwardly upward. The sum of my experiences on this dive was that I had been particularly impressed by the towering pillars of bundled kelp, and by the dark brown canopy, shot through by sunbeams. The power of the swells reaching me at ninety feet struck me also. The visibility was not bad either, about thirty to forty feet.

After lunch, we continued our drive south and headed for Jade Cove, still on the Big Sur coast. Jade Cove is not misnamed; real nephrite jade can be found there.

At Jade Cove we parked at the side of Route 1 and looked across an ice-plant meadow. The sea was well below the meadow, separated from it by a steep cliff. A trail had been cut into the cliff, a series of switchbacks, and wooden stairs descended the steepest places. The trail led down to a short crescent beach, where two old men were excavating holes below the high-tide line, looking for jade pebbles. There were prominent signs, visible to everyone who visited the park, warning against the taking of any minerals, presumably jade, from above the high-water line.

We suited up beside the car and hiked across the meadow to the lip of the cliff, followed the steep and winding trail down its face. The water entry was easy, a simple walk down the beach into the water. But the difficulty begins in the water; the cove contains at least six kinds of kelp, and swimming in it is like swimming in a salad. The water was jade green, too. Moderate swells moved the dense mat of mixed kelp species. We pushed through and worked our way into deeper water, where we descended to about fifteen feet (four-and-a-half meters) and began sifting through the bottom pebbles. I picked up something

that looked green and shiny and showed it to my partner. He shook his head. I looked some more. I showed him another find. He again shook his head. Meanwhile, he seemed to be accumulating a collection of pebbles of his own, but his were keepers. They shone with a deep, glowing green and could not be scratched by a steel knife point. Eventually I did manage to find a handful of very small pebbles of real jade.

The strong surge from the surface buffeted us about. I had to grasp a handful of kelp in order to stay in one place. My companion motioned me to an opening between two huge boulders; he went in and beckoned me to follow. The cave was open at both ends, and I could see greenish daylight at the far opening, fifteen feet away. My partner began scraping with a knife at one of the walls of the narrow chamber, then grasped the flashlight that had been hanging loosely at his wrist. He beamed the light on the rock. I moved in closer to see what held his interest. He pointed with his knife. The hard steel blade could not scratch the smooth surface. I borrowed his light and shone it on the rock; it was a dark watery green and as slick as glass. Its color had a depth that drew my eye beneath its surface. Could the entire giant boulder, as big as a small house and weighing many tons, be solid jade? In fact, massive chunks of jade have been found here, and in 1971 a 9,000-pounder was recovered by a dedicated group of divers. While I had been in the cave, the light at the far opening was quite beautiful, a small triangle of bright green. The effect was also quite striking when I was just outside the cave, looking in at the yellow glow from my companion's lamp. We signaled each other almost simultaneously that we were running low on air. We moved into shallow water, surfaced through the salad, and waddled up the beach. We then faced the task of climbing up the trail that clung to the cliff.

Inside the Inside Passage

The northwestern coast of the United States and the southeast coast of Canada together comprise a region that takes its special character from a unique combination of circumstances. The special feature of this rugged stretch of coastline is an enormous inland waterway, the Inside Passage, which cuts across national boundaries and includes Puget Sound in the United States and the Strait of Georgia in Canada. Although the Inside Passage continues to the north, ultimately reaching Skagway, Alaska, it is the more limited waterway of Puget Sound and the Strait of Georgia that is of special interest to the cold-water diver, because this waterway provides matchless diving within reach of the population centers of the western North Pacific.

This waterway opens to the Pacific broadly at the Strait of Juan de Fuca (the strip of water separating western Washington from British Columbia) and somewhat more diffusely through the island-choked Queen Charlotte Strait, at the upper end of Vancouver Island. Although the exposed coasts of Washington and Vancouver Island experience some of the heaviest surf in the world, the sheltered nature of this inland waterway permits little wave action here, except in the Strait of Juan de Fuca, which opens directly to the Pacific. The waves that do occur in these protected waters are primarily of local origin.

This great waterway is hemmed in by mountains. Thus, the Strait of Georgia nestles between the mountainous Vancouver Island on one side and the coastal range of mainland British Columbia on the other; and Puget Sound is bounded to the east by the Cascades and to the west by the coast ranges, including the Olympic Mountains, the Willapa Hills, and the mountains of Vancouver Island, which are an extension of these coast ranges. The snow-capped Cascades are part of the so-called ring of fire, the chain of volcanic mountain ranges that nearly encircles the Pacific. The intense mountain building that has taken place in this area has, by raising great crustal blocks above the general level of the continent, caused widespread subsidence that has brought flooding of river-cut or glacier-cut valleys and troughs as the sea invaded the falling land. This irregular coastline is an example of a drowned topography.

Puget Sound is the 100-mile long submerged northern end of a 350-mile trough whose southern end is the Willamette River valley in Oregon. Though originally river-cut, the northern end of this great trough was gouged by glaciers into a complex of deep, steep-sided channels; it is an enormous glacially carved scar. As recently as 15,000 years ago this region was overlaid by several thousand feet of ice; therefore, much of the bottom of Puget Sound is glacially deposited sand, clay, or gravel. However, the more than 450 islets and rocks that comprise the San Juan Islands, lying just south and east of the international boundary separating the United States and Canada, are outcrops of ancient bedrock that were not buried by glacial deposits, and hence provide substrate and shelter for a rich and varied hard-bottom community. The distance of these islands from the developed areas of Canada and the United States is responsible for their wild, virginal character. Swift tidal currents, up to seven knots, scour some of the passes between the islands. Puget Sound was explored by the British navigator George Vancouver in 1792; today, its shores are rimmed by a megalopolis made up of the contiguous cities of Seattle, Everett, Tacoma, and Olympia, the state capital.

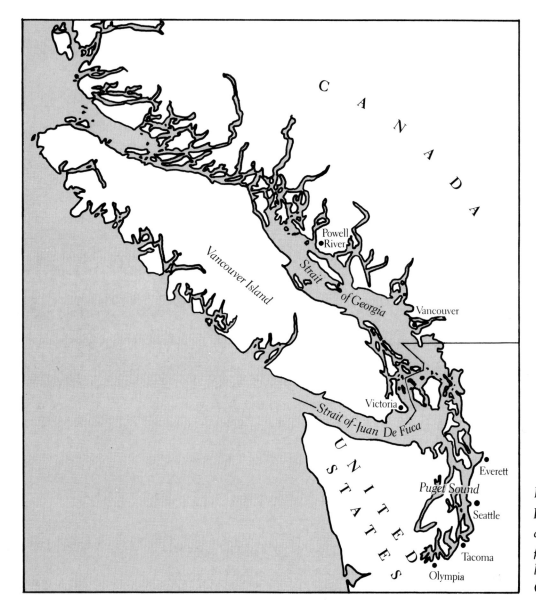

Figure 4.2
Puget Sound and the Strait of Georgia
are parts of the Inside Passage, a great
protected waterway that cuts across the
boundary between the United States and
Canada.

BENEATH COLD SEAS

The Strait of Georgia is more than 140 miles (225 kilometers) long and has a midchannel depth ranging from 900 to 1,200 feet (275 to 370 meters), but there are deeper submerged valleys. At its southern end, the Strait of Georgia is straddled by the Canadian cities of Victoria (the provincial capital, on Vancouver Island) and Vancouver (on mainland British Columbia).

The Strait of Juan de Fuca is the third arm of this great waterway. If you are willing to accept some geographic distortion for the sake of clarity, you can visualize the Strait of Juan de Fuca as the upright of the letter T, with the Strait of Georgia and Puget Sound each forming half the letter's crosspiece, the Strait of Georgia to the left and Puget Sound to the right. In reality the T would be lying on its side and the crosspiece would be quite lopsided, since the Strait of Georgia is approximately twice as long as Puget Sound. The Strait of Juan de Fuca opens to the Pacific and is bordered to the south by the majestic Olympic Mountains. This strait continues out under the Pacific as a great submarine canyon, and like the Monterey and Carmel submarine canyons to the south, provides an enormous reservoir of fertile, deep-ocean water that is carried into the surface waters of the strait by upwelling. These enriched surface waters are carried throughout much of the rest of the waterway by powerful tidal currents that can exceed five knots at Tacoma Narrows, in southern Puget Sound. To convey an idea of the flushing effect of these tidal currents, it has been estimated that 5 percent of the waters of Puget Sound are exchanged with each change of the tide. The flushing due to tidal currents in the Strait of Georgia is equally impressive. Because of the influx of nutrients carried into the waterway by these tidal currents, the marine life in the more favorable locations can grow to a truly remarkable size. A diver in these waters can find enormous starfish, sea anemones, chitons, and fleshy sea pens (91).

Puget Sound

I flew to Seattle, where I was met by a generous host. That night I slept on the waters of Puget Sound aboard my host's boat, and next morning we motored down the sound to Tacoma Narrows. The narrows are the gateway to the southern sound, and strong currents are common here.

We made our first dive in the narrows on the wreck of the old Tacoma Narrows Bridge, near slack water. We made a drift dive, with another friend following overhead in the boat, watching our bubbles. I was impressed by the numerous giant barnacles that studded the rocks; some were six inches in diameter at the base (79). They all had two sharp beaks and featherlike cirri. They seemed to grow on every available hard surface, even on other barnacles. The bottom was gravel and boulders with occasional exposed banks of clay that were scoured by the powerful current. Rockfish flew by in the swift current. I saw a colorful sun star with a pastel blue star superimposed on a pale orange body; it was *Solaster stimpson*. My companion surprised a big octopus, about three feet in diameter, and he wrestled with it for a while; but in the strong current it was hard to maintain a stable posture on the bottom. I braced my feet against rocks and held on to handfuls of short kelp. The best strategy was to kneel on the bottom, facing the current, with the sole of each foot pressed against a boulder.

My diving companion held the octopus' head, and I let

some of the animal's suction disks touch my glove. Although it grasped with just a few disks near the tip of one tentacle, the octopus was able to create enough suction to nearly pull my glove off when I drew my hand away. Its gill inlets and outlets inflated and deflated alternately as it moved water over its internal gills. The skin at the back of its sacklike head was wonderfully warty, and this bulbous sack tapered to a graceful point at the rear. It had the same beguiling habit of curling the tips of its tentacles into tight spirals that I had observed in the much smaller individuals I had encountered before. We released it and it jetted away, inking twice to persuade us that the two large brown blobs were actually the octopus. I pushed off the bottom and sailed over it in the swift current.

A huge concrete block loomed ahead. A pair of them had been used to moor the vessels working on the new bridge during its construction. The strong current nearly blew me around the corner of one of these massive cubes, and I had to hold on hard to keep from being carried off. These blocks were decorated with barnacles and anemones. Buffalo sculpins, warty, boxlike fish, lurked in the crevices under the blocks.

Near the end of the dive, I was enveloped in a dense school of fingerlings, a seemingly endless, writhing ribbon of fish that extended from one horizon of visibility to the other. Individual fish in the school occasionally rolled, flashing their silvery sides. I advanced toward the wall and it parted, as if by magic. It looked as if the fish would have to hit me, but none did. The giant ribbon divided ahead of me and closed behind, completely encircling me. I was cut off from my companion—and from everything else, it seemed. This was the first time that my underwater visibility had been curtailed by a wall of fish. I recalled that in California, off *Truth*, I had experienced lowered

visibility because of a great density of living organisms, but they had been tiny, nearly microscopic. I no longer had visual contact with my diving partner. I was running low on air and had already started to use my reserve breathing air, so I was not able to search further. I ascended alone. At the surface I found my partner waiting for me about twenty yards away. Our friend had followed our bubbles and had piloted the boat to within several yards of us. I swam to the side of the boat, removed my weight belt and scuba and passed them up, and hauled myself aboard. Then we helped my diving companion aboard.

After a surface interval of one hour, we moved the boat to Titlow Beach. Here I dived around an old ferry landing, mostly a stockade of pilings. Near the remains of this old landing rests a sunken wooden barge. Underwater, the pilings were covered with white and brown sea anemones, as was the wooden barge. In fact, there were so many anemones that I could see no wood on the pilings. It was a spectacular sight; some of the anemones were in full bloom, their tentacles fully extended. Others were semiretracted and limp, and yet others were contracted into tight balls. Small starry flounders, *Platicthys stellatus*, lay still on the sand, their upper sides speckled with a pattern of white spots that recalled the night sky. Spiny red sea cucumbers were everywhere there was crumbled wood. The presence of wood crumbs from the old landing suggested that this bottom was sheltered from the strong current that scoured the main channel of the southern sound.

I carefully watched a large sunflower star for many minutes. Among the fluffy tufts on its back it had a veinlike network of a strikingly intense orange; I had noted this before in other sunflower stars. I flipped it onto its back and watched it carefully as it righted itself. It curled opposing arms together to form a cylinder, then did a slow cartwheel. The first few times I saw this righting behavior, in other waters, it had always appeared that the star had gotten on its "feet" again without my being able to understand how; I thought I understood it this time, but I could not begin to explain such a complicated maneuver.

On the sunken barge I saw two nudibranchs on patches of bare wood; one was pale yellow, its back studded with orange-tipped papillae; it was *Triopha carpenteri*, the orange-tipped nudibranch. The other was all white, or possibly a very light yellow, with no cerata, but it did have a rosette of gills on its back and two horns on its head; this was the white sea lemon, *Archidoris odhneri*. It retracted both its horns and the circle of gills when I stroked its back. A spiral coil of nudibranch eggs lay near the first nudibranch (100). Ribbed brown kelp and bull kelp anchored themselves to occasional hard spots on the bottom. I surfaced to find the boat waiting.

My next dive was off the public beach in Edmonds, Washington, in northern Puget Sound. Two sunken steel dry docks and a submerged wooden barge provide shelter and substrate for an abundant variety of marine life, and this area has been specifically designated a state underwater park to protect its marine life.

We entered the water from a small sandy beach next to an active ferry landing. We waded out to thigh-deep water, then snorkeled about a hundred yards across a rippled sand bottom on which we saw countless small sanddabs. When the water reached five feet of depth, there was a sharp line where a zone of eelgrass, *Zostera marina*, began. This sea grass continued to a depth of ten or twelve feet. Small delicate white sea anemones adhered to the sea grass blades; often their pedal disks wrapped around the edges of a blade. One individual overlapped two blades, holding them together like crossed swords. Many tiny jellyfish, perhaps an inch in diameter, moved through the sea grass. Their transparent bells revealed a quartet of striking orange internal organs, and their bells were delicately fringed. This diminutive animal was the orange-striped jellyfish, *Gonionemus vertens*.

We finned beyond the zone of sea grass and glided into a grove of mixed kelp, mostly ribbed *Costaria* and bull kelp. A current swept the bull kelp into compound angles: the slender stipes leaned with the current, whereas the tassels surmounting the single bladder streamed nearly horizontally.

We soon reached a very picturesque steel wreck, one of the dry docks. It lay in twenty-five feet (nearly 8 meters) of water and was carpeted—no, upholstered—with anemones, most of which were white; others were light pink, yellow, and brown. Heavy-bodied box crabs and decorator crabs, with their borrowed finery of sponges and other encrusting organisms, hung on to the steel plates where the anemones left them a little space. Beneath the wreck was a haven for colorful rockfish, sullen buffalo sculpins, and flounders with periscopic eyes. I peered further under the wreck, and a huge and very ugly sculpin peered back.

In several places holes were rusted clear through the double hull; I looked up through such an opening toward the surface and I beheld a beautiful vista through the dark frame: waving kelp and milling rockfish against the wind-roughened surface mirror. In a burst of exuberance I swam through the hole.

NORTH OF CONCEPTION

Strait of Georgia

I arrived at Vancouver airport at around 11:00 A.M. and cleared Canadian customs with no more than a slap on my scuba cylinder. A good Canadian friend was waiting. We drove from the airport directly to Brunswick Beach, Vancouver, for my first dive in the Strait of Georgia. The coarse gravel berm was littered with giant logs at the storm-tide line. The beach is on Howe Sound, an inlet off the Strait of Georgia. There was negligible current and no wave action. The air temperature was a very comfortable 80°F (27°C); the water temperature turned out to be 60°F (16°C) at the surface and 50°F (10°C) below. We were quite comfortable in our full cold-water suits.

The bottom was unusual, resembling the bottom of a freshwater lake. It was covered with a fine, gray silt and seemed fairly barren and uninteresting. But that appearance was deceiving; I got the first inkling that this underwater environment was much more than it seemed when I saw a clump of very large white sea anemones on slender stalks attached to a large boulder. Their stalks were twelve inches or longer. Some were nearly upright, their tentacles fully extended in a perfect corona. Others had their tentacles involuted but their stalks were still upright, and yet others were rolled up like freshly laundered socks. I soon noticed that other rocks, too, supported clusters of these striking flowerlike animals.

At a depth of twenty feet (6 meters) I saw the syruplike effect of the incomplete mixing of fresh with salt water: an underwater spring or surface runoff was being mixed into the salt water of the sound.

As we dived along the steeply dropping bottom, we encountered a monumental boulder. We peered into a hollow underneath and saw an ugly gray face. It was a wolf eel, *Anarrhichtys ocellatus*. I pointed my finger at the face, and it pulled back into the hole. My partner signaled emphatically: "No. Danger." There was urgency in his gesture. Later, ashore, he estimated the fish could have been six feet long or more. The ugly snout did have an impressive set of teeth. I had seen fish in New England waters that we called wolf fish, and they had a similar appearance: a body color like lead, slate, or leather, an ugly, almost human face, and a long naked body tapering to a pointed tail. This was *Anarhichas lupus*, a different species, and I had never seen a specimen as large as this one.

I also saw a stand of broad-bladed ribbed kelp that had ridiculously small holdfasts, probably *Costaria costata*, and among their blades moved small dark fish with a patch of bright yellow on their backs and dorsal fins. These were quillback rockfish, *Sebastes maliger*. Purple sea stars (actually the ochre sea star, *Pisaster ochraceus*) aggregated into small groups. Other sea anemones, smaller than those that had struck me at the start of the dive, caught my eye. They had beautifully banded tentacles, and I named them peppermint anemones, although their common name is Christmas anemone, *Tealia crassicornis* (94). We saw hundreds of coon-stripe shrimp, *Pandalus danae*, whose red bodies are striped with blue (93). These tiny animals can shoot backwards with a flick of their tails. A semitransparent shape moved slowly in midwater; it was a sea walnut, a comb jelly or ctenophore. It had tiny glowing lines of beating filaments along the meridians of its ellipsoidal body, and these flickered as this tiny creature rowed itself through the water.

We had seen a small collection of pebbles and other colored

148

objects near the entrance to the wolf eel's lair. This was an octopus garden, gathered and placed there by an octopus, so my companion said. But no octopus was to be found. My diving companion said later that he felt certain that the wolf eel, a newcomer to the boulder, had eaten the octopus and then taken over its lodgings as well. We saw several other smaller octopus gardens, each in front of a hole under a boulder. But we saw no octopus.

Powell River

The drive from Vancouver to Powell River, British Columbia, is punctuated by two ferry rides: one bridges Howe Sound from Horseshoe Bay to Langdale, the other furrowed the waters of Jervis Inlet from Earls Cove to Saltery Bay. The big car ferries are clean and well maintained, and these trips across the water saved us a considerable amount of travel, compared to driving around the shores of Howe Sound and Jervis Inlet. Between ferry rides, we drove along the Sechelt (rhymes with *seashell*) Peninsula, on a highway that skirted the Strait of Georgia.

At Powell River we established our base of operations at a hotel that specialized in diving, Beach Gardens Resort. There we were given directions to a local wooden wreck that promised to provide an interesting dive. The hulk of the *Malahat* was once a proud sailing vessel; it was later converted to a barge, and finally sunk.

The water was warmed by the sun at the surface, but it was quite cool just a few feet below, perhaps 55°F (13°C). We snorkeled out along a breakwater of riprap until we reached a conspicuous spot of white that the resort's diving director had painted on the breakwater. The spot marked the beginning of a compass course that led to the bow of the wreck in thirty feet of water; the stern lay in seventy feet. We found the old boat exactly where it was supposed to be. The decaying wooden hull was stripped, with little to offer in the way of artifacts for the souvenir hunter; but more importantly, it provided a haven for marine life, including several thirty-inch lingcod, and many quillbacked rockfish and large sunflower stars (80). I also saw a quite large white nudibranch, two to three inches long. Much of the bottom was covered with the large ribbed kelp with very small holdfasts that I had seen off Brunswick Beach. Many shrimp lived on the bottom among the kelp. As we finned back to shore along the base of the breakwater, we saw many rockfish, lingcod, and, in the shallows, clusters of beautiful purple sea stars.

Rock Walls

Our next dive was to be a drift dive, always a delightful luxury. We descended next to a vertical wall of granite at a place called Iron Mines. At a depth of about sixty feet in this very clear water we first saw large hollow boot sponges; even larger cloud sponges began to occur at about eighty feet. At ninety feet I saw a boot sponge with a tree leaf resting in it. Rockfish darted across the face of the wall: tiger rockfish with vertical stripes, copper rockfish with a single horizontal stripe, and quillback rockfish with a striking patch of white on their backs. A large Puget Sound king crab, *Lopholithodes mandtii* (92), was nearly concealed among the sponges. It appeared deep burgundy in the available light, but when I switched on my underwater lamp, the king crab was revealed to be a brilliant rust red. Numerous small quillback rockfish, galathoid crabs (like miniature lobsters), and orange coon-stripe shrimp all took refuge within the cloud and boot sponges. The cloud sponges had interconnected cavities; these unusual sponges were white and quite beautiful. Numerous nearly spherical tennis-ball sponges, *Tetilla arb*, looked exactly as if some careless player had lobbed them onto this submerged rock. We looked up and glimpsed a school of herring well above us; sleek and shadowy salmon darted rapaciously into their midst.

Below 60 feet (18 meters) the lateral visibility was quite good; at 100 feet (30 meters) it was excellent. We could see approximately 80 feet (24 meters) ahead. A swift current swept us along the wall. The resort's diving director was up top in the rubber boat, moving to keep above our exhaust bubbles; he was within a few feet of us when we broke the surface. We hauled ourselves on board and headed back to the hotel.

After a two-hour surface interval we made our second dive of the day on another wreck, a forty-two-foot-long wooden tug, the *D and D*, that had been deliberately sunk to provide a focal point for diving. The hull provided shelter for marine life on this otherwise unremarkable sand bottom. There were lingcod everywhere—some quite large—the biggest aggregation of these fish I had seen; as we approched, they waddled off. Rockfish abounded too, coppers and quillbacks. Two large sunflower stars, one approximately thirty inches in diameter, the other twenty-four inches, engaged in a confrontation. The larger one appeared to glide over the bottom like a stiff carpet. It moved surprisingly fast toward the smaller one; its tube feet moved in an interestingly coordinated fashion, as if attached to wheels. Was I witnessing aggression? I turned the bigger star on its back. It expended an agonizing effort to right itself; with its underside

exposed it instinctively knew the extreme danger it faced. The big star had some trouble righting itself on the fairly steep sand; it kept sliding down the slope, but eventually it rolled over.

In shallower water several chitons (the lined chiton, *Tonicella lineata*, with its jointed, shield-shaped carapace) and a single magnificent fallen archangel, the nudibranch *Dirona albonineata* (96), about three inches long, worked their way over the small rocks. The nudibranch's white, petallike cerata completely covered its body; it is also called the alabaster nudibranch.

Though sand bottoms are often fascinating, rock bottoms may harbor a greater variety of marine life. We found this to be true at a submerged rocky wall called Okeover Caves. Here the bottom was a sheer vertical wall of rock. Red rockfish played over the rugged surface, and numerous small moon jellies pumped themselves through the water. These jellyfish had a delicate fringe at their margins and four orange rings—internal organs—were visible through the tops of the translucent bells; and there were ruffled petticoats just underneath the bell. I also saw something I had never encountered before: huge sea blubbers, *Cyanea capillata*, another kind of jellyfish; they were at least eighteen inches in diameter with long cordlike tentacles. They pulsated hard to drive themselves quite slowly through the water.

There was a cave at ninety feet (twenty-seven meters) and its floor was littered with a pile of rocks, probably fallen from the ceiling; the ceiling itself was carpeted with white sea anemones. Here, on the pile of rocks, one of the sea blubbers had "crashed and burned," like some alien aircraft. The downed creature looked like a collapsed cream puff. Outside the cave, whole walls were covered with more white anemones, punctuated with translucent tunicates; brittle stars hung in crevices. I looked up the vertical wall and saw a solitary sea blubber silhouetted against the blue-green water.

After a surface interval of forty minutes, we dived off a beach. The steeply sloping bottom was gravel and sand interspersed with small boulders. Hundreds of moon jellies, two to four inches in diameter, filled the water, their white bells like the canopies of parachutes. The whole scene reminded me of an airborne assault in which thousands of paratroopers were floating down to earth. We encountered two octopuses and caught one. It promptly inked, producing a swirling burst of dark brown. The blob looked vaguely like an animal in the water, a clear demonstration of how the octopus uses its ink to elude a predator. I stroked the creature with a fingertip between its eyes to pacify it, as one diver I knew suggested. The octopus become still. Its pupils were slits and constricted even further. The octopus turned darker on my black mitt; it had been gray on the gravel. It tightly curled the tips of its tentacles in the same direction and seemed to fall asleep in my hand. What caused this animal to relax so completely? Was it trust or resignation, I asked myself. My mind drifted away to muse on the vulnerability of all that I had seen beneath cold waters: the ugly yet beautiful goosefish, the red-shirted garibaldis, the sleek halibut, and the octopus. That defenseless creature now limp in my grasp brought my thoughts back to the present, and I remembered that I had asked myself a question. Trust, I answered hopefully. I hoped that I deserved that trust. I hoped that we all deserved it.

NORTH OF CONCEPTION

INDEX